Anywhere Computing with Laptops

Making Mobile Easier

HAROLD DAVIS

DISCARDED

que®

800 East 96th Street,
Indianapolis, Indiana 46240

International Standard Book Number: 0-7897-3327-7

Library of Congress Catalog Card Number: 2004111339

Printed in the United States of America

First Printing: August 2005

08   07   06   05   4   3   2   1

Trademarks

All terms mentioned in this book that are known to be trademarks or service marks have been appropriately capitalized. Que Publishing cannot attest to the accuracy of this information. Use of a term in this book should not be regarded as affecting the validity of any trademark or service mark.

Intel, Intel Centrino, the Intel Centrino logo, the Intel logo, Pentium, and Intel SpeedStep are trademarks or registered trademarks of Intel Corporation or its subsidiaries in the United States and other countries.

Warning and Disclaimer

Every effort has been made to make this book as complete and as accurate as possible, but no warranty or fitness is implied. The information provided is on an "as is" basis. The author and the publisher shall have neither liability nor responsibility to any person or entity with respect to any loss or damages arising from the information contained in this book or from the use of the CD or programs accompanying it.

Bulk Sales

Que Publishing offers excellent discounts on this book when ordered in quantity for bulk purchases or special sales. For more information, please contact

U.S. Corporate and Government Sales

1-800-382-3419

corpsales@pearsontechgroup.com

For sales outside the United States, please contact

International Sales

international@pearsoned.com

ASSOCIATE PUBLISHER
Greg Wiegand

ACQUISITIONS EDITOR
Todd Green

DEVELOPMENT EDITOR
Kevin Howard

MANAGING EDITOR
Charlotte Clapp

PROJECT EDITOR
Tonya Simpson

COPY EDITOR
Rhonda Tinch-Mize

INDEXER
Ken Johnson

PROOFREADER
Elizabeth Scott

TECHNICAL EDITORS
Will Eatherman
Eric Griffith
Mark Reddin

PUBLISHING COORDINATOR
Sharry Lee Gregory

MULTIMEDIA DEVELOPER
Dan Scherf

DESIGNER
Anne Jones

PAGE LAYOUT
Brad Chinn

Contents at a Glance

Table of Contents

About the Author

Harold Davis is a strategic technology consultant, hands-on programmer, and the author of many well-known books. He has been a popular speaker at trade shows and conventions, giving presentations on topics ranging from digital photography through wireless networking, web services, and programming methodologies.

His books include several about wireless networking, including *Absolute Beginner's Guide to Wi-Fi* (Que) and *The Wi-Fi Experience: Wireless Networking with 802.11b* (Que).

Harold has served as a technology consultant for many important businesses, including investment funds, technology companies, and Fortune 500 corporations. In recent years, he has been Vice President of Strategic Development at YellowGiant Corporation, a company providing infrastructure for Internet marketing, Chief Technology Officer at a CRM analytics startup, a Technical Director at Vignette Corporation, a leader in customer-centric content management, and a Principal in the enterprise consulting practice at Informix Software, a leading database company. Harold's work at Informix focused on the CRM and EAI needs of major Informix customers, including Sprint, United Airlines, and Wells Fargo.

Harold started programming when he was a child. He has worked in many languages and environments and been lead programmer and/or architect in projects for many corporations, including Chase Manhattan Bank, Nike, and Viacom. His private enterprise consulting work has ranged from the highly technical to strategic market positioning for clients that ranged from startups to large enterprises.

He has earned a Bachelor's Degree in Computer Science and Mathematics from New York University and a Juris Doctorate from Rutgers Law School, where he was a member of the law review.

Harold lives with his wife, Phyllis Davis, who is also an author, and their three sons in the hills of Berkeley, California. In his spare time, he enjoys hiking, gardening, and collecting antique machines, including typewriters and calculation devices. He maintains a Wi-Fi Access Point and a mixed wired and wireless network for the Davis menagerie of computers, running almost every imaginable operating system.

Dedication

For Julian, Nicholas, and Mathew

Acknowledgments

Thanks to all those who helped make this book possible, including Todd Green and Matt Wagner. And a very special thanks to Chris Hopper and Phyllis Davis, who provided great quantities of energy, expertise, hard work, and technical savvy.

We Want to Hear from You!

As the reader of this book, *you* are our most important critic and commentator. We value your opinion and want to know what we're doing right, what we could do better, what areas you'd like to see us publish in, and any other words of wisdom you're willing to pass our way.

As an associate publisher for Que Publishing, I welcome your comments. You can email or write me directly to let me know what you did or didn't like about this book—as well as what we can do to make our books better.

Please note that I cannot help you with technical problems related to the topic of this book. We do have a User Services group, however, where I will forward specific technical questions related to the book.

When you write, please be sure to include this book's title and author as well as your name, email address, and phone number. I will carefully review your comments and share them with the author and editors who worked on the book.

Email: feedback@quepublishing.com

Mail: Greg Wiegand
Associate Publisher
Que Publishing
800 East 96th Street
Indianapolis, IN 46240 USA

For more information about this book or another Que title, visit our website at www.quepublishing.com. Type the ISBN (excluding hyphens) or the title of a book in the Search field to find the page you're looking for.

Introduction

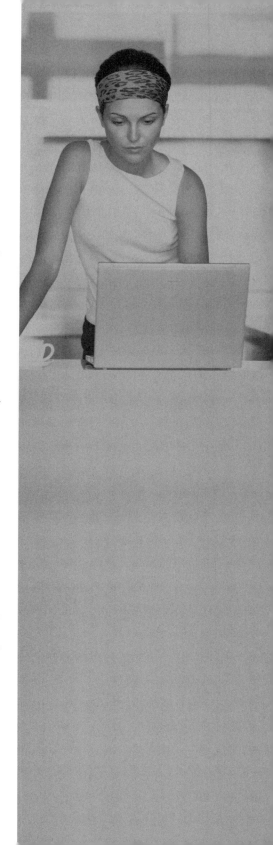

Have you ever wanted to lounge on a beach chair at a fancy resort and surf the Internet? Connect and get your email in a coffee shop such as Starbucks, or one inside a Borders bookstore? Get your email while waiting at an airport? Put together some computers in your home so that they can share files or access to the Internet without drilling holes or snaking snarled wires from one computer to another? Do your work using your laptop from your garden patio?

With your laptop that uses Intel Centrino mobile technology and a public hotspot that uses the wireless networking technology known as Wi-Fi, you can do all these things, and more.

This book shows you how.

I don't assume that you know anything about Wi-Fi, or about any of the related topics, such as how to set up a network of computers. You'll find everything you need to take your wireless computer on the road, and to set up a wireless network, right here between these pages. You'll also learn how to get the most out of your Centrino laptop and wireless technology in general.

So step right up and get ready to enter a wonderful new world without wires!

In this book, I hope you'll find inspiration as well as practical information. I believe that Wi-Fi wireless networking is a technology that has the power to have a huge and positive impact. You can harness that power simply with your Centrino laptop.

This is wonderful material, and it's a lot of fun! So what are you waiting for? It's time to Wi-Fi with your Centrino laptop!

How This Book Is Organized

Anywhere Computing with Laptops: Making Mobile Easier is organized into five main parts, as follows:

- **Part I, "Mobile Computing Quick Start,"** is a general introduction to the Intel Centrino mobile technology platform, wireless networking, Wi-Fi technology, and using Wi-Fi. If you want to know how to get your Centrino laptop up and running quickly with wireless technology, this part will get you there without delay.

- **Part II, "Getting the Most from Your Mobile Computer,"** tells you how to use software that is fun (and in some cases, very useful) to get more out of your laptop that uses Intel Centrino mobile technology. In this part, you'll learn how to become more productive with your Centrino laptop, take pictures with your laptop (if it is equipped with a camera) and transmit them over wireless, use your laptop as a telephone (with the right equipment), stream audio and video files over your wireless connection, play games, and more.

- **Part III, "Mobile Computing on the Road,"** shows you how to use your Centrino laptop on the road, explains the best road warrior tools to bring, teaches you how to find the best places to connect, and gives you tips on making the best deal with Wi-Fi service providers.

- **Part IV, "Your Own Wireless Network,"** explains everything you always wanted to know but were afraid to ask about successfully setting up and managing a wireless network in your home or small office.

- **Part V, "Securing Your Computer and Network,"** explains how to safely and securely deploy Wi-Fi on the road and at home.

Besides the 18 chapters in 5 parts in this book, I've also provided several appendixes that I think you'll find helpful:

- **Appendix A, "Wireless Standards,"** provides more details about the ins and outs of the 802.11 wireless standards.

- **Appendix B, "Where the Hotspots Are,"** shows you how to find public Wi-Fi hotspots and provides specific information (which is hard to find in one place) about Wi-Fi locations such as airports, hotels, and retail stores. If you travel with your Centrino laptop, you might find that this appendix alone is worth the price of the book!

- **Appendix C, "Intel Centrino Mobile Technology Platform,"** provides in-depth information about the components that make up Intel Centrino mobile technology—used in the bulk of laptops purchased today.

- **Appendix D** is a glossary. *Anywhere Computing with Laptops: Making Mobile Easier* lies at the intersection of a number of technologies—wireless broadcasting, networking, and personal computing. Each of these fields in its own right is replete with acronyms, jargon, and technical terminology. The glossary provided in Appendix D will help you hack your way through this morass of incomprehensible techno-babble and be your helpful companion through a benighted sea of obscurity.

Taken together, the 18 chapters and 4 appendixes in *Anywhere Computing with Laptops: Making Mobile Easier* provide all the information you need to successfully and happily use your Centrino laptop for mobile computing. In other words, this book is out to make mobile work for *you*.

Conventions Used in This Book

Although it is my hope that you can figure out everything in this book on your own without requiring an instruction manual, it makes sense to mention a couple of points about how information is presented in this book.

Web Addresses

There are tons of Web addresses in this book, mostly because these are places you can go for further information on a variety of related topics. Web addresses are denoted using a special font. For example,

```
http://www.wi-fiplanet.com
```

You'll find a complete electronic version of the book in PDF format on the CD-ROM included with this book. Each of the web addresses is clickable within the PDF file, allowing you to instantly access each website, provided your computer is connected to the Internet.

Special Elements

This book also includes a few special elements that provide additional information not in the basic text. These elements are designed to supplement the text to make your learning faster, easier, and more efficient.

NOTE

A *note* is designed to provide information that is generally useful but not specifically necessary for what you're doing at the moment. Some are like extended tips—interesting, but not essential.

TIP

A *tip* is a piece of advice—a little trick, actually—that lets you use your computer more effectively or maneuver around problems or limitations.

Let Me Know What You Think

I always love to hear from readers, particularly if they have nice things to say about my book! Seriously folks, I'd love to hear from you with comments, criticism, praise, and (most important) items that should be in the next edition of this book. Please write to me at AnywhereComputing@bearhome.com.

Although I can't promise to answer every email, I will do my best to do so.

Thanks for reading my book!

MOBILE COMPUTING QUICK START

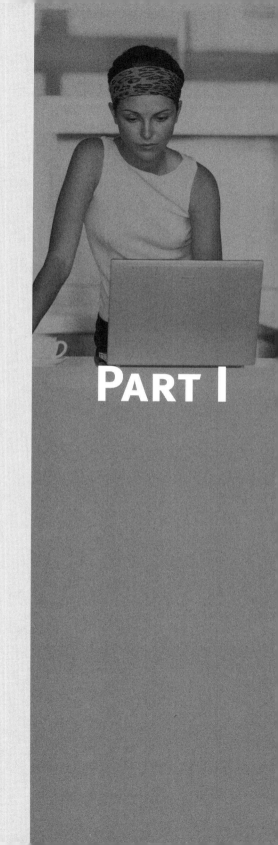

PART I

Understanding Intel Centrino Mobile Technology

What is "anywhere computing"? Anywhere computing means the ability to use your computer to connect safely and securely—anywhere, without wires! It also means that this is easy: Connecting to the Internet or your own private network anywhere should be no harder than simply turning on your computer.

This book shows you how you can practice anywhere computing.

Before you know it, you'll be using wireless networking—also called Wi-Fi (short for wireless fidelity)—to check your email and surf the Internet poolside at hotels, in airport waiting areas, and at coffee shops and restaurants. You'll also learn how to safely access the resources available on your own networks from these remote locations using Wi-Fi. And, you'll find out how to use wireless technology to set up home or small office networks so that you never again have to string wires (or drill holes in the wall)! In short, anywhere computing will empower mobility in your life.

Intel Centrino mobile technology is a great enabler of anywhere computing. This chapter explains the advantages of the technology and Intel's vision of the mobile future.

Intel Centrino Mobile Technology and Wi-Fi

Intel Centrino mobile technology is a platform incorporating Intel's technologies designed specifically for mobile computing with integrated wireless LAN capability. It is intended to enhance mobile performance and extend battery life when incorporated in notebook computers. But the main thing is that a Centrino laptop combines a central processing unit (CPU) and chipset designed for mobile computing with wireless LAN functionality using the Wi-Fi standards. (You'll find more information about Wi-Fi in Chapter 8, "Entering a World Without Wires," and Appendix A, "Wireless Standards.")

IN THIS CHAPTER

- Intel Centrino mobile technology and Wi-Fi

- The components of Intel Centrino mobile technology

- Changing the way people use computers

- Clearing bottlenecks and road-blocks

- The Intel wireless computing vision

- The future is faster; wireless is everywhere

Unlike many other computers that can be used with wireless networks, Centrino computers provide an integrated Intel Wi-Fi product, called *Intel PRO/Wireless Network Connection*. This Wi-Fi network connection works together with the other Centrino components to provide freedom and flexibility so that you can work and play at home or on the road without hunting for a phone jack or a network cable. Best of all, there's no need to plug in a special card because it's all built in.

The Intel PRO/Wireless Network Connection uses standard Wi-Fi protocols or "flavors" (which are described in detail in Appendix A). In fact, four different "flavors" of Wi-Fi are supported, depending on which Centrino computer you own or buy:

- The Intel PRO/Wireless 2100 Network Connection, supporting 802.11b
- The Intel PRO/Wireless 2100A Network Connection, which is dual band, supporting 802.11a and 802.11b
- The Intel PRO/Wireless 2200BG Network Connection, which is dual band, supporting 802.11b and 802.11g
- The Intel PRO/Wireless 2915ABG Network Connection, which is dual band and tri-mode, supporting 802.11a, 802.11b, and 802.11g

If you are deciding which mobile computer to buy, or if you are just getting started using your laptop for anywhere computing, your computer probably comes equipped with the Intel PRO/Wireless 2200BG Network Connection, which supports 802.11b and 802.11g.

It's also reassuring to know that Intel supports a variety of industry standard and leading third-party security solutions enabling safer notebook connectivity using Wi-Fi. Intel is also working with virtual private network (VPN) hardware and software vendors to verify security with VPNs. For more information about security and VPNs, see Part V, "Securing Your Computer and Network."

The platform provides software that enhances performance in several ways (see "The Importance of Performance" later in this chapter). This software also helps to manage Wi-Fi connectivity and to switch between access points using a system of multiple profiles (see Chapter 3, "Configuring Your Mobile Computer," for details).

Platform Architecture

Centrino technology consists of three components:

- The Intel Pentium M family of processors, designed for enhanced mobile, low-power performance
- The Intel 855 or Intel Mobile 915 Express family of chipsets, which controls memory and graphics and is designed to work well in mobile applications with the Pentium M

- The Intel PRO/Wireless Network Connection, which implements Wi-Fi WLAN connectivity, as I explained to you earlier in this chapter

From a conceptual viewpoint, the relationships between these components are shown in Figure 1.1.

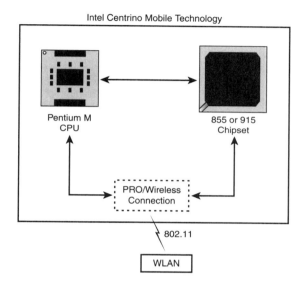

FIGURE 1.1

Intel Centrino mobile technology includes three major components.

The Benefits of Built-in Wi-Fi

Of course, Centrino technology is optimized for use with mobile computers. And it is built for wireless networking from the ground up—a great advantage over previous generations of mobile computers. There are no extra cards sticking out and no dangling antennas. Centrino-based laptops are sleek—the antennas are built right in. Bottom line, there is less stuff to carry around, wireless reception is great, your laptop is easier to use, and the battery lasts for hours.

Alternative Wi-Fi Solutions

If you use a mobile computer that does not incorporate Centrino technology, chances are it was not built from the ground up with Wi-Fi capabilities. This isn't the end of the world, but it means that a Wi-Fi "card" needs to be added to the mobile computer. The card might have been added at the factory, but if not, you'll need to configure it correctly. In either case, one of the expansion slots of your computer will be used. (So you can't use it for anything else.) In addition, the card might stick out of the side of the mobile computer, if it doesn't use an internal card, making its form factor more awkward. Figure 1.2 shows a notebook computer with an external Wi-Fi card seated in the expansion slot.

Although Centrino mobile computers are not the only ones to have internal, built-in Wi-Fi radios (rather than using an external PC-Card slot), they have the most tightly integrated processor and chipset architectures for enhanced performance and longer battery life.

FIGURE 1.2

On other mobile computers, to use wireless networking you have to add a Wi-Fi card in an expansion slot.

Most important of all, by choosing a Centrino laptop, you can be sure that you have a system designed from the ground up for anywhere mobile computing—one that has been extensively validated by Intel with hardware vendors of network access points and with hotspots around the world.

Mobile Computers Compared to Desktop Computers

Of course, you can connect desktop computers to wireless networks using Wi-Fi. This is usually done by adding a card to the desktop computer or by using an external Wi-Fi device connected to the computer using its USB connector. Either way, your desktop computers can now participate in the wonderful world of wireless, and you can say goodbye to drilling holes in your walls for network wires. (See Part IV, "Your Own Wireless Network," for complete information about creating your own wireless network.)

However, by definition, a desktop computer is not a mobile computer. Centrino technology is designed for low power systems and enhanced battery life, and generally to improve performance on the go. Desktop computers don't need these features.

The Importance of Performance

Intel designed Centrino technology to bring about a new era in mobile computing—incorporating fully enabled wireless networking, improved mobile performance, enhanced battery life, and thinner, lighter laptop designs.

Most people will agree that enabling a sleeker, thinner mobile computer system is a good and fun thing. I confess that I can remember the days when a mobile computer meant something so big and heavy that it took a weight lifter to manage it, and I sure am glad those days are gone.

But where the anywhere computing rubber really hits the road, so to speak, is in performance and enhanced battery life.

Performance

With throughput up to 54Mbps at 5GHz (802.11a) and 2.4GHz (802.11g), the Intel PRO/Wireless Network Connection enables fast network performance. (See Chapter 8 for more information about the different flavors of Wi-Fi and their relative speeds.)

In addition, the Intel Wireless Coexistence System helps reduce interference with some other kinds of wireless devices that may be competing with wireless networks.

Users may also engage the Power Save Protocol (PSP), a feature with five different power states, allowing power versus performance choices when in battery mode.

Enhanced Battery Life

The Intel PRO/Wireless Network Connection supports the extended battery life benefits of PSP. In addition, Intel Intelligent Scanning Technology reduces power needs by controlling the frequency of scanning for Wi-Fi access points.

Changing the Way People Use Computers

Perhaps you already use wireless networking, or perhaps you are just about to get started with it. Either way, you are one of a minority. Statistics show that only a fairly small percentage of people in the United States are using Wi-Fi. Less than 20% of households have some form of wireless networking in use, or have a wireless mobile computer. Turned around, this statistic means that more than 80% of us haven't yet experienced the joys of anywhere computing!

Why don't more people use wireless computing? There are a number of difficulties with wireless computing. Here are some of the most significant ones:

- It can be difficult to install. Only 28% of users in one survey were able to install a Wi-Fi card on their own.

- It can be hard to find the right wireless station (or SSID). This book will help you understand how to correctly find the SSID.

- Wireless computing inherently is less secure than wired computing.

- Wi-Fi coverage is incomplete, and access to wireless networks is not seamless.

Anywhere computing changes the way people live and work. We are no longer tethered to our fixed wires and locations. But before this freedom can be achieved, there is work to be done in clearing bottlenecks and roadblocks.

Clearing Bottlenecks and Roadblocks

Intel's goal is to help clear the bottlenecks and roadblocks that make using wireless networking more difficult than it needs to be. Centrino laptops help achieve this with

- Preinstalled Wi-Fi—you're ready to go.

- Built-in features to correctly find the right wireless station (SSID) to connect to. In Chapter 3, I'll show you how to set up your computer to take advantage of them.

- Support for all major wireless security initiatives. In Chapter 17, "Protecting Your Mobile Wi-Fi Computer," I'll explain how to practice safe mobile computing.

The good news is that the wireless computing from anywhere situation is getting better day by day. All the time, more and more access points are becoming available, connection speeds are gradually getting faster, and it is becoming much easier to use wireless networking (thanks, in part, to books such as this one).

I can't create a seamless national Wi-Fi network overnight by snapping my fingers, but I can show you how to configure your computer to best find wireless networks (Chapter 3), tips and techniques for finding wireless access points (Chapter 9, "Finding Hotspots"), and pointers on working with national Wi-Fi networks (Chapter 10, "Working with National Wi-Fi Networks").

Intel's Mobile Technology Vision

Intel Centrino mobile technology represents the goal of wireless broadband: It is always on wherever you are, it's speedy and fast, it's ready for use, and it's convenient to use.

With Centrino technology, you can use your computer from wherever you are to beam pictures to experts who can help explain them (see Chapter 5, "Taking Digital Pictures from Your Laptop," for more information about taking pictures with your Centrino computer). It means you can look up anything on the Internet whenever you need to. It means these things, and more.

This book will help to show you how to make the most out of your Centrino mobile computer as wireless computing exists today.

Intel's vision of the future is that mobile technology becomes totally transparent and easy to use. Networks mesh together and are self-configuring. Every computer is a network node, and all devices connect to each other in a seamless way. Security is not an issue. We can all work together from wherever we want, and no one is tethered to a cubicle.

The Future Is Faster; Wireless Is Everywhere

Today's laptops based on Intel Centrino technology offer a better mobile computing experience than ever before. But wireless networking innovation is not over. On the contrary, we are just getting started. Exciting new technologies are on the horizon that will offer even faster networking over longer distances. Here is a brief glimpse into what is
coming to mobile computers near you:

- 802.11n—An extension of existing 802.11 Wi-Fi technology to speeds of 100Mbps.

- WiMAX—A new technology based on the IEEE 802.16 with speeds of up to 70Mbps over distances of up to 30 miles.

- 3G (third-generation) CDMA—Cellular wireless data technologies available now in many U.S. metropolitan areas. Speeds vary from 100Kbps to more than 2Mbps depending on the cellular provider.

802.11n

802.11n is an evolutionary step from current Wi-Fi 802.11b/g/a technologies. 802.11n increases the speed of a local wireless network, enabling applications such as high-quality streaming multimedia. Although still in the final phases of becoming a standard at the time of this writing, Intel made public in June 2005 that it has produced a prototype chip combining 802.11n, b, g, and a standards in a single device. It stands to reason that the Centrino platform could incorporate this new chip in the future.

WiMAX

WiMAX is a new wireless technology (based on the IEEE 802.16 standard) that seeks to solve the "last mile" problem of bringing high-speed Internet service into homes and offices that are spread over a geographic area that is too great for economical installation of a wired network. WiMAX technology will operate at speeds up to 70Mbps over distances up to 30 miles. Intel introduced its first WiMAX product in April 2005 in the form of a chip aimed at manufacturers of base-station networking equipment. The Intel product roadmap indicates WiMAX technology could be incorporated into Centrino chipsets in 2006.

WiMAX has the potential for making seamless wireless access a reality. Its range means that you are much more likely to have a wireless access point available.

Intel is working with the wireless industry to spearhead the deployment of WiMAX. According to an Intel spokesman,

> "We as an industry are headed toward the 'broadband wireless era.' The next wave is about portability, with people wanting access anywhere. WiMAX will play a key role in delivering this."

Wi-Fi and WiMAX might prove to be more complementary technologies than competing technologies, with WiMAX supplying "last mile" capabilities to homes and offices and Wi-Fi provided connectivity inside your premises. However the two technologies are ultimately deployed, Wi-Fi is right here, right now. It makes no sense to wait for WiMAX when you can get started with Wi-Fi today.

3G CDMA Cellular Wireless Data

3G CDMA cellular wireless data services are not part of the family of Wi-Fi 802.11 technologies, and they are not part of the Centrino platform today, so these data services won't be discussed anywhere else in this book. They are mentioned here because they offer broadband-class Internet networking speeds right now in major U.S. metropolitan areas. Cellular providers such as Sprint and Verizon sell both the hardware (a card that plugs into a laptop) and the service to allow high-speed Internet access in most places you can use a cellular telephone. The hardware and service is expensive, but if there is a strong business need and you don't have the time to look for a wireless hotspot at Starbucks, 3G data services might be worth looking into.

Summary

Intel Centrino mobile technology is designed for wireless networking from the ground up. Centrino computers include built-in Wi-Fi wireless network access and don't need an additional wireless card. They are also optimized for mobile applications, with configurable power usage and battery-saving features.

Computers enabled with Centrino technology are ready to roll with Wi-Fi, but there are still some things you need to know, particularly about connecting to specific access points, security, and ease of use. (This book will help.)

The future will bring easy-to-use, completely seamless mobile computing available anywhere. Stay tuned!

Buying a Mobile Computer

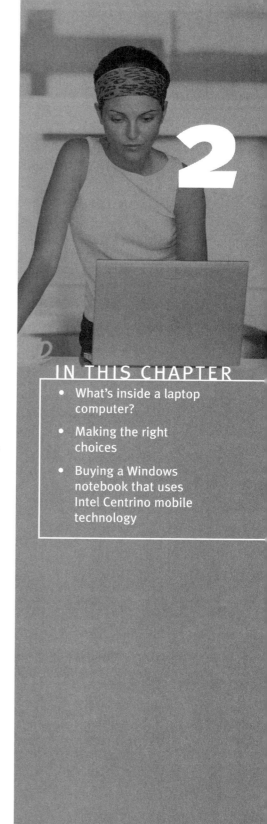

If you are reading this book, you are in one of two situations:

- You own a Windows-based wireless laptop and want to learn more about using its wireless features.

- You are considering purchasing a Windows-based wireless laptop.

One way or the other, you probably feel excited about mobile computing and are ready to go out and surf while you sip latte, read your email at the airport, or sit beside the swimming pool.

This chapter assumes that you are in the second group. You don't have a laptop that works with Wi-Fi, the networking technology that enables mobile computing. For more information about Wi-Fi, see Chapter 8, "Entering a World Without Wires."

So in this chapter, I suppose that you are getting ready to go out and buy a Wi-Fi mobile computer that runs Microsoft Windows and is based on the Intel Centrino mobile technology. You'll learn about the most important considerations when buying a mobile computer, things to look out for, what's really important, and the best places to buy your computer.

By the way, in this chapter, I am using the terms "mobile computer," "laptop computer," and "notebook computer" to mean exactly the same thing: In all cases, they refer to a very portable computer operating Windows and providing Wi-Fi wireless networking using Intel Centrino mobile technology.

For the record, and just so there's no misunderstanding, you have some other choices if you want to use Wi-Fi besides buying a new laptop with Intel Centrino mobile technology. First, there are wireless Windows laptops that don't use Centrino, and you can add a Wi-Fi card to almost any existing laptop. Second, laptops from Apple also use Wi-Fi. (In the Apple universe, Wi-Fi is called *Airport* and *Airport Extreme*.) Finally, it is even possible to use Wi-Fi with a laptop using Intel Centrino mobile technology running Linux (rather than Windows), although this is probably more appropriate for advanced users.

Of course, Wi-Fi also works well with desktop computers.

But from the viewpoint of this book, my assumption is that you either own (or are looking to buy) a Windows laptop based on the Intel Centrino mobile technology. In particular, this chapter is about buying a laptop to use for mobile computing.

Getting to Know Your Future Wi-Fi Laptop

Laptops work in pretty much the same way as full-sized desktop computers—they just come in a smaller package. Most everything is compressed into the small familiar form factor that you can carry around with you (unlike desktop computers, which typically feature separate display devices and system units).

So when you are learning about your future Mr. (or Ms.) Laptop Computer, you should know that (just like a desktop computer) your laptop will have

- A system unit (which includes the central processing unit, or CPU)
- A display device (laptop display devices are generally LCD, or liquid crystal display, screens)
- Peripheral devices, probably including a pointing device such as a trackpad or stick mouse that takes the place of a mouse and likely including speakers for sound

The laptop form factor typically includes the system unit, the display, and peripheral devices including a keyboard and pointing devices all in the single small, lightweight package. Essentially, these elements in the laptop are no different from the elements in a desktop computer; it is the small package size, also called the *form factor*, that makes a laptop computer what it is.

Because they are comparatively miniaturized and require some special engineering features (such as the capability to run on low power), laptops are more expensive than comparable desktops.

The system unit is the part of the computer that makes it a computer. Just like its big brother, the desktop computer, the system unit in your laptop has a number of important components, including

- A *microprocessor*, also called the *central processing unit* (CPU), which controls the entire computer.
- Short-term storage, called *random access memory* (RAM), which is used to temporarily store instructions and information that can be used by the microprocessor.

- Long-term storage, which is the hard disk used to permanently store important computer programs and data.

- Devices used to get information in and out of the computer; for example, CD drives, diskette drives, network cards, and Wi-Fi cards.

BATTERY LIFE AND LAPTOPS

Unlike desktop computers, mobile laptops are battery powered. Laptop computers provide a recharging mechanism for the computer battery when the computer is plugged in. All new laptops sold today come with Lithium Ion batteries—a change from the Nickel-Cadmium battery days.

How long a mobile computer can run on its battery is very important to users because this determines how long the computer can be used without plugging it in to an electric socket. The ability to work without network wires, thanks to Wi-Fi, is kind of undermined if you have to plug in to a standard electrical outlet just to get power.

Many factors go in to Lithium Ion battery life, including the power drawn by the CPU and the power needs of the computer's peripherals. This is an area to investigate carefully before you buy your laptop, based on your needs. So review battery life specifications carefully before you buy. On some models, multiple batteries are an option for extending usage time, so if this is important to you, you should investigate the feature before you buy.

The Pentium M, which is part of the Intel Centrino mobile technology platform, is specially designed to be used in laptops because it has low power draws. (Of course, the CPU isn't the only laptop component that draws power.) You can check to make sure that the laptop you are considering uses one of the microprocessors specially designed for laptops.

Here are some tips for prolonging the life of a Lithium Ion laptop battery:

- Make sure you plug in your laptop as often as possible. Unlike the older Nickel-Cadmium rechargeable batteries, Lithium Ion batteries do not have a *memory effect*. The memory effect, which occurred with Nickel-Cadmium batteries, resulted in a loss of battery capacity when the batteries were partially discharged and charged repeatedly. In Nickel-Cadmium batteries the memory effect could be overcome by one or more deep discharge/charge cycles. This was called *battery conditioning*. Battery conditioning is definitely not necessary with Lithium Ion batteries and it will not increase their capacity or capability to hold a charge. In fact, it will shorten the life of a Lithium Ion battery.

- Limit the number of full charge to full discharge cycles. A Lithium Ion battery has a maximum life of 500 full cycles. This means that if you use your laptop on the train in the morning and fully drain the battery, recharge it, and then fully drain the battery again in the afternoon you have already used two full cycles. If a battery has only 500 full cycles and you use two full cycles every day, you'll need to replace the battery within six to eight months! With partial discharges, a Lithium Ion battery can last as long as 1,000 cycles.

- Beware of fully discharging a Lithium Ion battery because the polarity within the battery cells could actually reverse and short circuit the battery.

It's likely that one of the things you'll focus on most when deciding which laptop to buy is the microprocessor. This makes sense because the speed of the CPU largely determines how fast the computer can perform operations.

NOTE
You shouldn't even judge a CPU just on the basis of its speed. Many facets of the architecture of a CPU besides its raw speed can affect its performance. For example, a Pentium M (powering a system using the Intel Centrino mobile technology) running at about 1GHz performs on par with an Intel Pentium 4 desktop computer running at about 2GHz.

With laptops, the CPU is particularly important. First of all, as I mentioned earlier, CPUs specially designed for use in laptops use less power—and therefore aren't as draining on battery life. Second, these microprocessors—such as the Pentium M that is part of the Intel Centrino mobile technology—don't run as hot as CPUs intended for desktops.

Just as brute brawn isn't everything in life, you probably would not be surprised to learn that the raw speed of the CPU isn't the only factor in how quickly a laptop performs its appointed tasks. For many applications, the amount of RAM available on the computer is actually more important than the CPU speed. In another example, for watching movies, besides the quality of the video screen, the most important hardware is the video display subsystem, not the CPU.

Trade-offs

A number of years ago, a friend of mine quipped that "the computer you really want always costs $5,000." Over time, the cost has come down, and you can certainly buy a high-end laptop for less than $2,000 today. But the point of the joke is still true. Unless money is absolutely no issue for you, you will have to make same trade-offs such as

- Faster CPU or longer battery life

- Lighter weight or less expensive

- Bigger and better display or less cost

For the most part, these choices will depend on your wallet. But a mobile laptop is a specialized computer, and some of the trade-offs really depend on how you will use the system.

For me, it is extremely important to have a lightweight, small machine, but I also wanted a reasonable size keyboard. I care more about the size, weight, and keyboard usability than anything else—including cost and screen size. So my choice in a mobile computer will be dictated by my preferences. The important thing here is to know yourself.

The general bottom line is do an assessment of what really matters to you, and purchase accordingly (see the following sidebar for more tips on this topic).

WHAT REALLY MATTERS

As I've noted, you'll have to make the final decision on what's really important to you in a mobile Wi-Fi laptop. Are you looking for a general-purpose mobile computer at modest price, or do you want a model with a really wide screen? Are you looking for the longest possible time operating on battery power, or do you need a lot of computing power for running complex programs?

To get another viewpoint on how to make this decision, I asked a friend of mine who is an expert consultant and has advised thousands of computer purchasers for her words of wisdom. Here's what the expert says:

- Buy a well-known name brand, such as Dell, IBM, Toshiba, HP/Compaq, or IBM. (You can certainly buy good equipment made by other vendors, but these four are among the most consistently reliable.)

- Buy new. If you are trying to choose between a used laptop and a new one, buy new if you can afford it. Mobile computing technology is advancing quickly. This year's new laptop is a lot better than last year's model, especially when it comes to LCD screen technology—current generation laptop displays are brighter and sharper and have more vibrant colors. Also, laptops are easily damaged. A used system might have been dropped or bumped hard enough to damage critical components such as the hard drive. However, there is no way to tell from the outside. To make a long story short, buy new.

- Don't be too cheap. You can expect a good piece of equipment to last a long time, so buy one that is rugged and with enough power.

- It's fine to buy a cute computer, but you don't need to fall in love with it at first site. In other words, don't buy a computer just because it has the cutest form factor.

- Buy a Centrino-based laptop. Look for the Intel Centrino logo (see Figure 2.1); it will be a sticker on the palm-rest or on the manufacturer's literature. The battery life improvements made possible by Centrino (up to four hours on a single charge with many models, as many as six hours with others) is strong enough justification on its own. Then, add the convenience and performance of built-in Wi-Fi and your decision is a no brainer.

FIGURE 2.1

When buying a laptop, look for the Intel Centrino logo.

- Ergonomics are important. Buy a model with a screen you like to look at and a keyboard and pointing device (track pad or stick mouse) that are comfortable for you to use.

Buying a Mobile Computer

When you first walk into a computer store or go online to a laptop manufacturer's website, you might be overwhelmed by the number of laptops from which to choose, and you might be confused because from the outside they look largely the same. Add to this the dizzying amount of mumbo jumbo when looking at the technical specifications, it can make you reach for an aspirin. But take heart! This section gives you a basic understanding of a few broad categories into which most laptops fall. With this information you can quickly cut through all the technical gunk and focus on the capabilities that really distinguish one laptop from another. With this understanding you will be able to focus on just the models that meet your needs and know this is the right laptop for you just by looking at the specs.

Let's begin to break down the laptops that are out there into categories that will make it easier for you to sort out the type that's right for you. The three major categories are

- Value priced
- High computing performance
- Ultra-long battery life

There are other product categories than these three. For example, laptops that are ultra compact, laptops that are exceptionally rugged with cases made of exotic materials, laptops that can be used as writing tablets, and so on. However, these types of mobile computers are not what most readers of this book need or are looking for.

Value-Priced Laptop Computers

This category offers good basic features at a reasonable price. These laptops are easy to spot on just price alone; they're generally between $1,000 and $1,400. Value-priced machines will meet the needs of most buyers, so unless you know you need high computing performance or ultra-long battery life, concentrate on products in this category. Don't worry; you will have plenty of models to choose from.

High Computing Performance

Laptops in this category have the fastest Intel mobile processors, use the highest-speed memory technology, and have powerful video chips. This type of laptop is well suited for people who run complex spreadsheets or perform some other task that is highly compute-intensive. The most advanced PC-based video games often require this type of power. So, if you are a financial analyst or a serious PC gamer, you might want to have a look at high-computing performance systems. Laptops in this category range in price from $2,000 to $3,000.

Ultra-Long Battery Life

The latest generation of Centrino technology based on the Intel Mobile 915 Express chipset now offers support for low-voltage (LV) and ultra low voltage (ULV) Pentium M CPUs. LV and ULV processors are extremely frugal when it comes to battery power and can go for six hours or more on a single battery charge. But expect to pay top dollar—laptops in this category range in price from $1,500 to $3,000.

Value-Priced Laptops: What You Need to Know When You're Ready to Buy

When buying a value-priced laptop, use the following list as a guideline for what to select. If you can afford more, great! But, make sure the laptop you get has at least these features.

- Processor—Buy a laptop with an Intel Pentium M processor. Intel offers a lower-performance processor called the Celeron, but I don't recommend it. Laptops built with a Celeron can appear enticing because of teaser prices starting at around $600, but the money you save in the short term is not a good long-term tradeoff. The Celeron processor does not run large complex programs as well as the Intel Pentium M does, and the Celeron does not have the Intel SpeedStep technology for conserving battery power. Therefore, the Celeron is slower and uses more power, so you get less battery life. In the years to come, software will offer more sophisticated capabilities and will demand more and more processor speed. The Intel Pentium M will keep pace better with new software technology because the underlying performance is there. And with a Pentium M you will be able to run longer on battery power. So spend a little more and get an Intel Pentium M.

- Memory—512MB DDR SDRAM (Don't get less!). 512MB is mandatory these days for good performance. Some manufacturers configure value priced laptops with only 256MB of RAM. This is just not enough. It takes almost 256MB just to run Windows XP Home and a good virus scanning product. If your laptop only has 256MB, there will not be enough RAM to run programs efficiently. Your laptop performance will be really slow. Make very sure you get 512MB. There is no need to get more than 512MB for just normal activities. If you are doing video editing, or some other task that needs a lot or memory, you might consider more. But that would be the exception and not the rule for most people. Also, upgrading RAM in laptops later is a tricky bit of hardware installation even for the most experienced user.

- Wireless—Intel PRO/Wireless 2200 internal wireless card (802.11b/g).

- Internal hard disk—30GB or larger. 30GB is enough for most uses and is the minimum size on an internal hard drive in most laptops today. Unless you are downloading thousands of songs for your Apple iPod, a 30GB internal disk drive should be enough.

- Optical disk—CD-RW/DVD-ROM combo drive (burns CDs, reads DVDs). I recommend a CD-RW/DVD-ROM drive because it is fun to watch a DVD on your laptop when the airline movie is horrible. CD-RW (Compact Disk Read/Write) drives are very convenient for making backups of your data and for exchanging files with others. There is even a CD burning wizard built in to Windows XP to make the process of making your own CDs easy. Also, the cost savings in getting a CD-only drive is very small, so get the CD-RW/DVD-ROM combo drive.

- Operating system—Windows XP Home. The big difference between Windows XP Home and Windows XP Professional is the capability to log in to a Windows domain. XP Home can't, XP Professional can. This is of concern to you only if you want to use your laptop in an office environment where the network administrator uses a Windows domain server to enforce security policies. The average user at home or school will be very happy with Windows XP Home.

- Productivity suite—Microsoft Office Basic (Word, Excel, Outlook). Microsoft Office Basic is the minimum recommended because some day someone will send you a Word or Excel document and expect you to be able to open it. The software that comes installed on your laptop is highly discounted, so there is no cheaper way to get Microsoft Office. Instead, you can get other flavors of Microsoft Office that contain additional applications such as PowerPoint for presentation graphics. Feel free to pay more to get these other applications if you think you will use them. You will never have a cheaper opportunity.

One item not in this list (because it is not consistently offered) is a high-capacity battery. Some laptop models have them, some don't. High-capacity batteries mean your laptop runs longer on a charge. And because laptop batteries degrade over time (fact of life, a laptop battery is considered a consumable) you will probably be able to go a year or two longer before you need to replace it if you get the high-capacity battery. This option can add as much as $130 to the purchase price but will give you many more hours to linger over a mocha at Starbucks instead of looking for tables near power outlets to plug in your charger.

What You Need to Know about Buying a High Computing Performance Laptop

So, you have a bit more money to spend and you want to impress your friends with the bone-crushing power of your mobile computer. Or you are a corporate executive with only hours to analyze complex numerical data before presenting it to the board of directors. Either way, you want all the computing performance you can get, and price is a secondary concern. Here's what you should look for when spec'ing a high-performance laptop:

- Processor—Intel Pentium M, 2GHz or faster. The fastest Intel Pentium M processors top out at 2.13GHz as of this writing. You should get at least a 2GHz processor. Faster is always better, but you can pay a real premium for that extra 10% of performance.

- Chipset—Mobile Intel 915PM Express (preferred) or Mobile Intel 915GM Express. Okay, get ready for the techno speak. In most cases I saved you the complication of technical details, but here it matters. The Mobile Intel 915PM Express or the Mobile Intel 915GM chipsets are specified because they support a 533MHz front side bus (FSB) between the processor and the memory. In layman's terms, a 533MHz FSB makes your computer run faster because data is exchanged between the processor and memory at a faster rate. The 915PM is preferred because it does not have a built-in graphics processor. This is desirable because Intel's graphics chips are somewhat lackluster compared to manufacturers such as ATI and nVidia. So the lack of a built-in graphics processor will mean the laptop manufacturer got a graphics chip from somewhere else, and it's probably better than Intel's.

- Graphics processor from a major graphics chip manufacturer such as ATI or nVidia—Keep in mind that if you are interested in game playing, you might want to check your favorite games to see what graphics they specify. That way, you'll get the processor you need.

- Memory—1GB DDR2 SDRAM. There are two things you should know. I specify 1GB here because you will probably run complex programs that need a lot of memory. Also, DDR2 memory is specified because it is the fastest laptop memory available.

- Wireless—Intel PRO/Wireless 2915ABG internal wireless card (802.11a/b/g).

- Internal hard disk—60GB or larger.

- Optical disk—DVD-RW/CD-RW combo drive (burns DVDs and CDs).

- Operating system—Windows XP Professional.

- Productivity suite—Microsoft Office Professional (Word, Excel, PowerPoint, Access, and Outlook).

What to Know when Buying an Ultra Long Battery Life Laptop

If you frequently find yourself on flights from New York to Singapore, you might want to consider a laptop in the ultra long battery life category. This type of laptop offers the absolute best that money can buy when it comes to long battery life. However, be forewarned that long battery life comes at a price. Processor speeds will be slower, to conserve power (even slower than a value-priced laptop), and you will pay a premium to get longer battery running time. Sorry, but there is no option today that combines high CPU performance and long battery life.

Here are the characteristics you should look for when buying an ultra long battery life laptop:

- Processor—Intel Pentium M Low Voltage (LV) or Ultra Low Voltage (ULV). As already mentioned, the Intel Pentium M LV and ULV processor run slower to conserve battery life. The top speed for this type of processor is 1.5GHz for LV and 1.2GHz for ULV. These speeds are still fast enough to get most work done, so if you need a long battery life laptop, you are not giving up that much when it comes to performance for most tasks. If you need to be absolutely certain, check with the laptop manufacturer or retailer to see whether they will accept a return if the laptop doesn't meet the performance requirements of your type of work.

- Chipset—Mobile Intel 915 Express family (915GM, 915GMS, or 915PM).

- Memory—512MB DDR or DDR2 SDRAM.

- Wireless—Intel PRO/Wireless 2915ABG internal wireless card (802.11a/b/g).

- Internal hard disk—40GB or larger.

- Optical disk—DVD-ROM/CD-RW combo drive (burns CDs, plays DVDs).

- Battery—High capacity, or room for a second battery in the laptop body. As mentioned earlier, if the manufacturer offers a high-capacity battery be sure to get one. After all, long battery life is the point here. In addition, with some ultra long battery life laptops you can remove the optical drive and put a second battery in the drive bay. This is very desirable because you get the benefit of double the battery life without having to shut down your laptop to change batteries in the middle of your working session.

- Operating system— Windows XP Professional.

- Productivity suite—Microsoft Office Professional (Word, Excel, PowerPoint, Access, and Outlook).

General Advice Regardless of the Type of Laptop You Buy

There are a few things you should keep in mind when buying any laptop. First, be sure you get virus protection. I will say this three times so it sinks in: Get virus protection, get virus protection, get virus protection! It doesn't matter whether you take the option from the laptop manufacturer or go out and buy it separately. Just get it. And keep the subscription up to date. A virus can really screw up your laptop and be really hard to get rid of. In extreme cases you will have to migrate your personal files somewhere else, then erase the hard drive and reinstall all your programs. A good computer consultant will charge between $200 and $500 for this procedure, depending on how much stuff you have on your hard drive. Don't take the risk. Get protected. And just like your laptop, pay for a good quality product from a major manufacturer. Symantec and McAfee are examples. Free virus-scanning products might seem attractive, but think about it: If the virus scanner is free, how do they pay their programmers to keep it up to date? There is no free lunch. Quality costs, so spend the money up front instead of having to pay a computer consultant down the road.

In addition, all laptops from major manufacturers offer a built-in modem, a built-in Ethernet interface, and a built-in sound card with internal speaker. Because these items are almost always included and are not major points of difference it is not worth discussing them.

Laptop manufactures also offer extra cost warranty, damage protection (such as cell phone insurance), and support services. Pricing for these extra services varies widely. Use your judgment as to whether they are right for you.

Where to Buy

There are numerous options about where to buy your laptop. You can purchase a machine

- Directly from the manufacturer, using the Internet or a catalog and telephone

- Online or using a catalog from a computer reseller

- At a "real" retail store such as a computer manufacturer's store, a consumer electronics store, a computer store, or an office supply store

There's nothing wrong with any of these options. It all depends on what you are most comfortable with and what works best for you. It's even okay to mix and match: You can check equipment out in a brick-and-mortar store, and then go ahead and buy it online.

Do make sure that you buy your Wi-Fi laptop from a reputable vendor who will make things right if there is a problem. It's a fact that the majority of hardware problems occur within the first twenty-four hours of use, so you want to be sure that the matter will be easily taken care of if you do happen to buy a lemon.

> **TIP**
>
> The first thing you should do when you get your laptop is run it for 24 hours. This should disclose most latent problems in the hardware. Some of this running time should be with the laptop on battery power (discharging), and some should be spent recharging with the laptop plugged in to an electrical socket.

You might be offered a floor sample, demo model, or reconditioned model (or the same thing by some other name) at a very substantial discount. This is a place for the old saying *caveat emptor*, "Buyer Beware!" to come into play. I'd recommend that you avoid any such scenario in which the laptop you are thinking of buying has been used and turned off and on, even if it is reconditioned (or factory certified—whatever that means).

It's easy to go to an online manufacturer's site to buy your Wi-Fi computer. For example, if you choose the Compare All Notebooks option on http://www.dell.com, you'll see the entire Dell current laptop offerings, as shown in Figure 2.2.

As shown in Figure 2.2, if you look at the models shown after you've clicked Home and Home Office, and then click Compare All Notebooks, in the middle you'll see two that are configured with Centrino-compliant technology from the get-go. These models are both designated with an "m" suffix (the Inspiron 600m and 700m). Other Dell models, such as the wide-screen Inspiron 6000, require you to select the options to make it Centrino compliant (a Pentium M processor instead of Celeron M).

Another thing to notice about the Dell website is that the ultra-long battery life models don't show up in the comparison view. You must select Latitude X1 from the list to see the models with the Intel Pentium M low-voltage (LV) and ultra low-voltage (ULV) processors.

FIGURE 2.2

By choosing Compare All Notebooks on the Dell site, you can get information about all the currently available laptops.

On http://www.ibm.com, click on Notebooks in the Shop for column. This will take you to the Notebook finder, shown in Figure 2.3. This is an excellent way to sort through the available models.

By clicking the Wi-Fi link shown at the bottom of the Select a Feature box shown in Figure 2.3, you will be presented with a feature selection list for all the models that are Wi-Fi–enabled (see Figure 2.4).

With your choices restricted to models that are Wi-Fi–enabled, you can now assemble a system that otherwise meets your needs. Because there are many choices of IBM laptops that are Wi-Fi enabled—Figure 2.4 shows thousands—it's a good idea to restrict your choice next by using some other criteria, such as the price range you are interested in or by specifying a particular CPU.

If you want to compare Wi-Fi notebook computers in general, a good resource to use is http://www.cnet.com. You can then enter a search term such as "Intel Centrino Notebook." The results will provide a great deal of pricing and review information about a wide variety of products (see Figure 2.5).

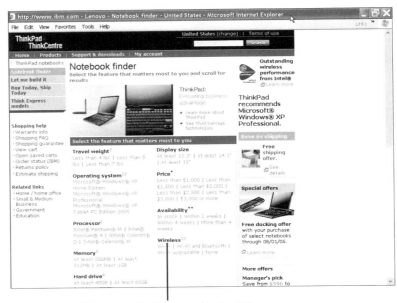

Click to see Wi-Fi–enabled laptop PCs

FIGURE 2.3

IBM's Notebook finder helps you sort through models.

From within CNET, you can generally click links to specific retailers to buy online at the prices quoted in the CNET comparisons.

A final point is that simply entering a phrase such as "Intel Centrino Notebook" into a Web search engine such as Google or Yahoo! will produce a great many links for researching wireless equipment and purchasing wireless notebooks. You should be aware that some of these links have been bought as advertisements, so you should not regard them as an endorsement.

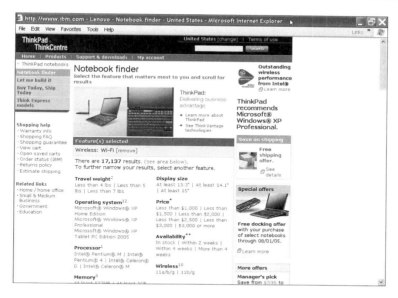

FIGURE 2.4

By selecting Wi-Fi, you can sort through all the models that are Wi-Fi–enabled.

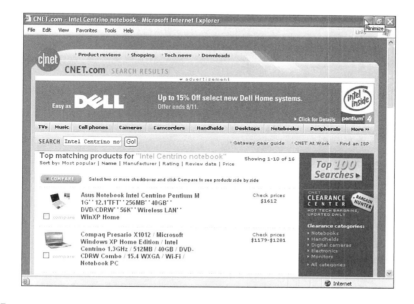

FIGURE 2.5

CNET provides an easy way to compare multiple brands of wireless notebooks.

Summary

Here are the key points to remember from this chapter:

- Buying a mobile computer can seem daunting, but it's not really.

- Determine for yourself what you care about most.

- Worry about weight, ruggedness, and battery life in a laptop, not cuteness.

- You can buy a Windows mobile notebook from a variety of sources including the manufacturer.

- It's very convenient to shop for a mobile laptop online, although you might want to visit a store to handle one before you buy it.

- Wi-Fi is the technology used to enable mobile computing.

- You might need to read specifications with some care, but you should be able to tell if a notebook computer has integrated Wi-Fi. If you are planning to use your laptop for mobile computing, don't buy one that doesn't come with Wi-Fi already onboard.

- Laptops based on the Intel Centrino mobile technology come pre-equipped with Wi-Fi.

Configuring Your Mobile Computer

This chapter explains how to use Windows XP to set your notebook computer based on the Intel Centrino mobile technology to work with wireless networks. Your mobile computer might possibly work with a wireless network without you having to do anything. However, it is also possible that you will need to tweak some simple settings, such as telling your mobile computer which wireless signal to connect to (if your computer detects more than one).

Some other things might need to be set. For example, if your wireless network is running with encryption turned on (as I advise in Part IV, "Your Own Wireless Network"), and you don't have a server to supply the encryption key automatically (as most small office or home networks don't), you'll need to know how to enter an encryption key for the wireless network. Also, you might need to know how to choose between different wireless signals to pick the right one.

This chapter explains everything you need to know to use Wi-Fi to connect your laptop based on Intel Centrino mobile technology to a home or small office network.

Working with the Control Panel

"All roads lead to Rome," as they used to say, and in Windows XP, there are many ways to get to the wizards and dialogs that are used to set networking and wireless connectivity options. So don't get confused just because there is more than one way to access these dialogs. It's okay whichever way you get to them. In this chapter, I'll show you how to always navigate to these dialogs—but don't worry if you open them another way on your particular Centrino laptop.

Depending on your mobile computer, it's quite likely that you can open these dialogs by double-clicking an icon on the Windows XP task tray that stands for your wireless connection. The appearance of this icon will vary depending on the make of your computer.

If you are using your mobile computer to access a public wireless hotspot—such as the T-Mobile hotspots at your local Starbucks—you might not need to worry about much of this configuration stuff. In Chapter 8, "Entering a World Without Wires," I explain the procedure for connecting to a T-Mobile hotspot (which is essentially similar to other public Wi-Fi providers).

I'll go over this in more detail in Chapter 8, but in a nutshell, you first have to set the SSID (or network name). At Starbucks, it is set to tmobile, the general name for T-Mobile's wireless hotspots. This SSID (and the SSID of all available networks) should appear in the Wireless Network Connection Properties dialog automatically, and all that remains to do is select it. By the way, the easiest way to see the wireless networks that are available is to right-click the network device in the Network Connections window or on the Windows taskbar and choose View Available Wireless Networks.

Next, you open a Web browser. If you already have a T-Mobile account, you can enter your ID and password. If you don't have a T-Mobile account, this is where you arrange to pay for your wireless access as I explain in Chapter 8. (If paying for wireless access in public places bugs you as it does me, see the section about finding free hotspots in Chapter 9, "Finding Hotspots.")

NOTE The description and figures in this chapter show how to accomplish tasks with mobile computers running Windows XP with Service Pack 2 installed. If you are running a version of Windows XP prior to Service Pack 2, some screen images will look slightly different but the procedural steps are largely the same. If you would like to update your version of Windows XP by installing Service Pack 2, click the Start button, and then choose All Programs, Windows Update. In the Browser Update dialog box that opens, click Express to download Service Pack 2.

But whatever the make of your computer, you can access the wizards and dialogs that are used to configure networking and access to wireless networks via the Windows Control Panel. So it's a good idea to learn how to set these things using the Control Panel in the first place.

Sure, you might learn how to open the Wireless Network Connection Properties window using the wireless configuration utility that shipped with your computer, and it might be easier to access this window from the icon on the Windows taskbar than using the Control Panel.

After you've opened the Wireless Network Connection Properties window, the functionality the window provides works the same no matter what route you took to open it.

But to back up for a minute, let me show you how to get help with some common networking tasks.

A foolproof way to get to the dialogs that manage networking—and wireless networking—is to use the Windows Control Panel. You can open the Control Panel by clicking the Windows XP Start button and then selecting Control Panel from the right side of the Windows Start menu.

When the Control Panel opens, it will show you a list of Control Panel items that can be used to configure your computer. This list is shown in Figure 3.1 in so-called Details Classic mode.

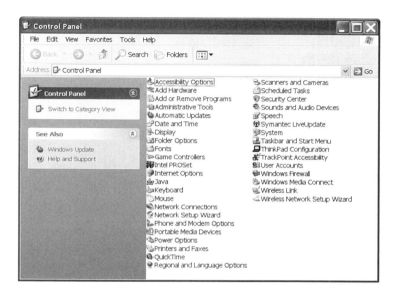

FIGURE 3.1

The Control Panel provides access to the utilities that help with network connections and configuration.

Don't let the various possible view modes in the Control Panel throw you. Similar to most windows that you see in Windows Explorer, the way you view the configuration utilities shown in the Control Panel can be configured. For example, instead of the Classic view shown in Figure 3.1, items can be categorized. And, in Classic view, the utilities can be shown as thumbnails or icons. (The view setting is customized using the View menu in the Control Panel.) So the first time you open the Control Panel, the items listed might differ in appearance and layout from those shown in Figure 3.1. Also, the precise items you see will depend on your computer (although some items are common to all computers) .

Help with Network Tasks

Everything having to do with network settings—including wireless network settings—is controlled by the Network Connections dialog, shown in Figure 3.2. This dialog will provide you with help with almost all common network tasks.

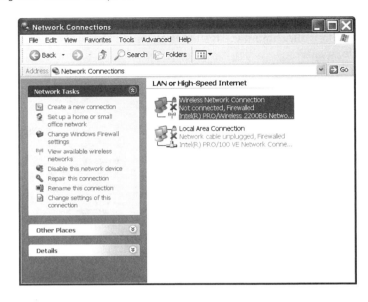

FIGURE 3.2

The Network Connections dialog is used to configure network settings, including wireless networks.

To open the Network Connections dialog, double-click Network Connections in the Control Panel.

If you look at the Network Connections window (shown in Figure 3.2), you'll see a Network Tasks pane at its upper left. The items in this pane will vary depending on how your Centrino system is configured. In almost every case, you'll see Create a New Connection and Set Up a Home or Small Office Network links. Clicking these links will start wizards—including the New Connection Wizard and the Network Setup Wizard—that will walk you through the tasks involved.

> **TIP**
>
> If the Control Panel is in Category View rather than Classic View, first click on the Network and Internet Connections link. Next, choose Network Connections from the Control Panel icons.

Working with the New Connection Wizard

To open the New Connection Wizard, click Create a New Connection in the Network Tasks pane of the Network Connections window. When you click this link, a screen welcoming you to the wizard, shown in Figure 3.3, will open.

FIGURE 3.3

The New Connection Wizard's opening screen shows that it is intended to help you with a variety of tasks related to network connectivity.

Click Next to get started with the wizard. The Network Connection Type pane, shown in Figure 3.4, will open.

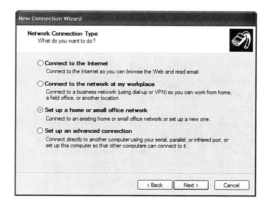

FIGURE 3.4

The Network Connection Type pane of the New Connection Wizard lets you choose the kind of connectivity task you want help with.

As the subtitle of the Network Connection Type pane of the wizard says, this is the point in the Wizard in which you tell it what "you want to do." The choices are

- Connect directly to the Internet (see Chapter 8 for information about connecting directly to the Internet using wireless technology)

- Connect to the Internet at your workplace using a virtual private network (VPN) or dial-up mechanism (see Chapter 17, "Protecting Your Wireless Computer," for information about connecting to a VPN)

- Set up a home or small office network

- Set up an advanced connection. This choice enables you to connect to another computer directly for file sharing or it enables your computer to accept connectins from other computers on the Internet, in effect acting as a VPN server.

If you guessed "door number three," or "Set up a home or small office network," you are probably right. This is the branch of the wizard that is most likely to get you started at your home or small office with connecting your Centrino laptop to a wireless network.

 TIP

If you don't have any wired computers and just want to connect a Centrino laptop to a wireless access point (the access point would itself probably be connected to the Internet via cable or DSL), see Chapter 11, "Networking Without Wires," for help and guidance. This chapter makes the basic assumption that you probably already have some sort of network going.

So make sure that Set Up a Home or Small Office Network is selected (it is the third option from the top as shown in Figure 3.4), and click Next. The Completing The New Connection Wizard pane will open. Click Finish. The New Connection Wizard will close, and then the Network Setup Wizard will open.

Using the Network Setup Wizard

The New Connection Wizard described in "Working with the New Connection Wizard" dovetails seamlessly into the Network Setup Wizard. It is here in the Network Setup Wizard that you will make choices that set several important Windows network settings, including

- How your mobile computer will connect to the Internet

- How your mobile computer will be named on the network

- Whether to enable printer and file sharing

- Turning on Windows Firewall

Doing the same tasks manually would require you to go to four different places to set Windows network parameters. The Network Connection Wizard enables you to specify all the settings in just one place.

If you are moving straight through this chapter of the book, you arrived at this section because you ran the New Connection Wizard described earlier. If not, there are two easy shortcuts you can use to get to the Network Setup Wizard. Here is the first way:

1. Click the Start button.

2. Select the Control Panel.

3. In the Control Panel select Network Setup Wizard.

Here is the second way:

1. Click the Start Button.

2. Select Control Panel.

3. Choose Network Connections.

4. Select the Setup a Home or Small Office Network option located on the left side of the Network Connections window.

Regardless of how you got there, the first screen of the Network Setup Wizard is a friendly welcome message that summarizes what the wizard will do. When you see the Welcome to the Network Setup Wizard pane, click Next to continue.

The Before You Continue pane opens. Here you are asked to review a checklist for creating a network, to install network cards and cables, to turn on all computers, and to make sure you are connected to the Internet. Perhaps you've noted the references in this list to things such as network cards and cables. However, with a Centrino laptop, we don't need no stinkin' wires! So ignore the references to old-style hardware, and boldly and bravely click the Next button.

Now you might see a panel informing you that the wizard found disconnected network hardware. This is likely to be the case if your Centrino laptop has a network card intended for use with a wired network (in addition to its wireless capability). The right strategy here is once more to ignore that which doesn't apply. Be sure the Ignore Disconnected Network Hardware box is checked, and click Next to continue.

What you see next depends on how your network is configured. If you are just starting out and there is no device on your network that already has a connection to the Internet (or if Windows can't detect it), you will see the Select a Connection Method pane shown in Figure 3.5. If this is the case, select the middle option—This Computer Connects to the Internet Through a Residential Gateway or Through Another Computer—and then click Next.

If you did not see the Select a Connection Method pane, you will see the Do You Want to Use the Shared Connection pane in Figure 3.6. In this case, select Yes, Use The Existing Shared Connection, and then click Next.

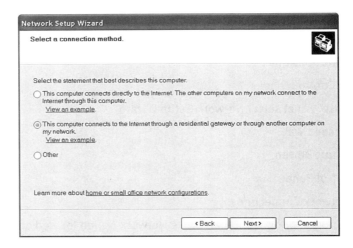

FIGURE 3.5

The Network Setup Wizard will display this pane if it cannot detect an existing shared connection to the Internet.

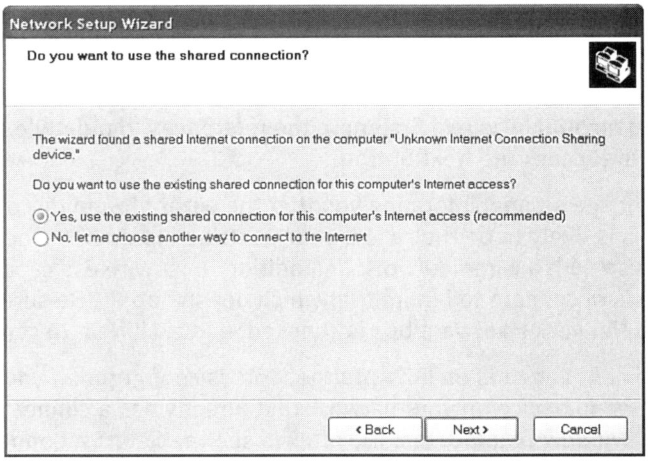

FIGURE 3.6

The Network Setup Wizard will display this pane if a shared Internet connection is found.

The next step is shown in Figure 3.7. In the Give This Computer a Description and Name pane you will specify how your computer is identified (just as you would if you were using a regular, old wired network). Give the computer a name, along with an optional description. This is your chance to replace a cryptic name that might have been assigned by your laptop manufacturer, such as ARCX10347AB, with something a bit more personal and descriptive. Click Next to continue.

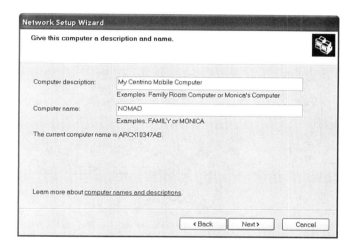

FIGURE 3.7

Give your computer a meaningful and unique name on the network.

In the next wizard pane, you will be asked to specify your workgroup name. All the computers, wired and unwired, that are part of your network should most likely be part of the same workgroup, so that workgroup name is what should be entered here.

If you don't know what workgroup name to use, you can use an existing computer on your network to find out. On the computer that is already connected to your network, click the Windows Start button. Next, highlight My Computer. Right-click My Computer, and choose Properties from the context menu. The System Properties window will open. Click the Computer Name tab. The name of the workgroup that the computer is part of will be displayed about midway down the Computer Name tab.

With the workgroup name entered, click Next.

You will now be asked for your file and printer sharing preference. I recommend you turn off file and printer sharing for now. This is the most secure way for you to start out. Don't worry; you can easily enable file and printer sharing later if you need to. Click Next to continue.

The wizard now displays a summary of the information you've entered so far, as you can see in Figure 3.8.

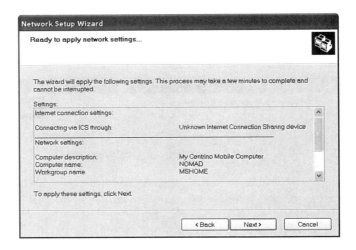

FIGURE 3.8

The wizard gives you a chance to review the computer name and workgroup settings before applying them.

The next pane displayed by the wizard states—perhaps a little optimistically—that "You're almost done..." On this pane, you are asked to choose from four options:

- Create a Network Setup Disk

- Use the Network Setup Disk I Already Have

- Use My Windows XP CD

- Just Finish the Wizard. I Don't Need to Run the Wizard on Other Computers.

The assumption behind this list of choices is that now that you've set up one computer on your network, you might need to use the same settings for other computers on the network. But our task has been to simply add a computer to an existing network; we don't need to change the settings on any other computer.

Choose Just Finish the Wizard, and click Next. The wizard will inform you that you have successfully set the computer up for networking and suggest that you might want to change the settings for file and folder sharing. It's important to know that the way you have sharing set up has security implications. For some sharing considerations and suggestions, and for information about how to change these settings, see Chapter 18, "Securing Your Wi-Fi Network."

Click Finish to close the wizard. If you are informed you need to reboot your computer, be sure all your files are saved and your applications are closed, and then click Yes to restart now. You probably won't be surprised to note that although we have successfully named our computer, and (in theory) added it to a workgroup, we don't have wireless connectivity yet.

I'll show you how to engage wireless connectivity with your Centrino laptop in the next section, "Understanding the Network Connections Window."

Understanding the Network Connections Window

If you've run either the New Connection Wizard or the Network Setup Wizard for the first time, it is likely that at this point your computer has a name, and an identity as a member of a Windows workgroup, but is still not connected wirelessly.

To fix this, have another look at the Network Connections window, shown again in Figure 3.9.

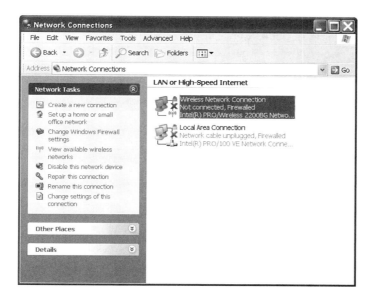

FIGURE 3.9

The Network Connections window shows that the wireless network is not enabled.

The right pane of the Network Connections window, marked LAN or High-Speed Internet, lets you know that the wireless connection is not working. You can tell this in Figure 3.9 because there is a big, bad red X right through the representation of the wireless Centrino hardware.

Right-clicking the icon representing the nonworking wireless adapter brings up a context menu with a variety of useful items, including Enable and Status. As you'd expect, Enable lets you enable and disable wireless networking. Selecting Status brings up a status window that provides information on the performance of your wireless network.

> **NOTE**
> Many laptops provide an external switch that enables or disables wireless networking. This is a great feature because it can be used to save power when you don't need wireless networking, and it makes it easy to turn wireless networking off in certain situations (for example, boarding an airplane). However, if your laptop is equipped with this kind of switch, you do need to make sure that it is turned on.

But the most generally useful item on the context menu is Properties, which opens the Wireless Network Connections Properties window. As you'll see in the next section, using the Wireless Network Connection Properties window, you can configure your wireless adapter in almost any way you can imagine.

Using Wireless Network Connections Properties

As I've explained in this chapter, the Wireless Network Connections Properties window is opened by first opening the Network Connections item from the Windows Control Panel. With the Network Connections window open, highlight a wireless connection from the right pane. Right-click to open the context menu associated with the wireless connection, and choose Properties to open the Wireless Network Connection Properties window, shown in Figure 3.10.

FIGURE 3.10

The Wireless Network Connection Properties window is used to configure wireless connections.

Essentially, the Wireless Network Connection Properties window is the command center for a specific wireless adapter, such as the wireless capabilities "baked into" a laptop with Intel Centrino mobile technology.

When the Wireless Network Connection Properties window opens, the General tab will be displayed as you can see in Figure 3.8. To get started configuring your wireless connection, click Wireless Networks. The Wireless Networks tab will be displayed, as shown in Figure 3.11.

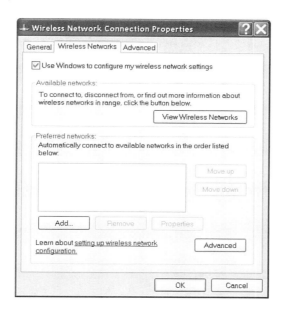

FIGURE 3.11

The Wireless Networks tab of the Wireless Network Connection Properties window is the command center for configuring wireless connections.

At the top of the Wireless Networks tab, you'll see the Use Windows to Configure My Wireless Network Settings check box. To make sure that the windows, wizards, and dialogs explained in this chapter work the way I've explained, this box should be checked. If it is not, wireless network settings might end up being configured using utilities provided by laptop vendors such as IBM, or using software such as Intel PROSet/wireless, provided to some vendors such as Dell.

The most important interface on the Wireless Networks tab is the Available Networks area, shown toward the top of Figure 3.11. Click the View Wireless Networks button to see the wireless networks you can connect to.

I've chosen to emphasize using Windows to configure wireless network settings because that way everything works in a standard way, and I can be sure that it will work on your Windows XP laptop that uses Intel Centrino mobile technology no matter what its brand. But if you choose to use a vendor's utility—called a device management utility, or DMU—rather than Windows XP to configure your wireless settings, you should know that it will set more or less the same things. Depending on your needs and preferences, you might even find that the alternative DMU is easier to use than Windows.

When you click the View Wireless Networks button, your computer scans the area for Wi-Fi signals and displays the wireless networks it finds in the Choose a Wireless Network dialog shown in Figure 3.12. If you want to rescan for wireless signals, perhaps because you've moved your laptop to a new location, click the Refresh Network List command under Network Tasks on the top left of the window. The strength of the signal from each wireless network is indicated by green bars. Just like a cell phone, the more bars, the stronger the signal.

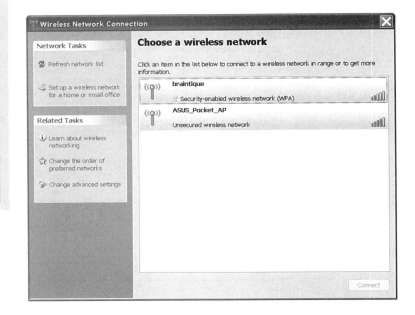

FIGURE 3.12

Choose a Wireless Network shows all wireless networks within range and indicates signal strength with green bars.

Wireless networks are listed by *SSID, service set identifier.* The SSID is used to identify the "station" broadcasting a Wi-Fi signal. The SSID is also sometimes called a *network name* or a *wireless network name.* (Apple calls the SSIDs used by its AirPort wireless products its *AirPort ID.*)

You should know that some wireless stations (also called Access Points or APs) do not broadcast their SSID (usually for reasons of security). There's nothing intrinsic to stop you from connecting to one of these networks, but you need to know the SSID—it won't show up in the Choose a Wireless Network list no matter how many times you click Refresh. For more information about why a wireless network would operate with this kind of stealth, see Chapter 18.

Connecting to a Wireless Network

To connect to a wireless network for the first time, select the network by SSID in the Choose a Wireless Network list. Next, click the Connect button in the lower right.

Depending on the security settings of the wireless network, you will get one of a couple different types of warnings after you click the Connect button. Notice in Figure 3.12 that the wireless network called Braintique is security-enabled while the wireless network called ASUS_Pocket_AP is not. When you connect to Braintique you see a message requesting a network key as shown in Figure 3.13. You need to enter a special key code to complete the connection. This key would be given to you by the person who administers the security-enabled wireless network.

> **TIP**
> If you are using a utility provided by your hardware vendor rather than the Windows XP wireless configuration software explained in this chapter, it might have a button that explicitly scans for wireless signals (functionally, much the same as the Refresh Network List command, but explicitly labeled Scan).

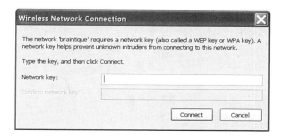

FIGURE 3.13

Security-enabled wireless networks require a network key to establish the connection.

The wireless network called ASUS_Pocket_AP is unsecured, which means there is no security restriction—anyone could connect to this wireless network. Because unsecured wireless networks (also called open access points) have no security restrictions, there is a possibility that someone else could "listen in" on the data sent and received by anyone who connects. That's why Windows gives the warning shown in Figure 3.14. Clicking the Connect Anyway button establishes the connection to the unsecured network. Most networks in public places (coffee shops, hotel lobbies, public libraries, and so on) are unsecured, so you are likely to see this warning when you are out and about. You learn more information on the possible risks of connecting to an unsecured wireless network in Chapter 18.

Congratulations, you're connected! Teacher gives you a gold star! Regardless of whether the wireless network is unsecured or security-enabled, once connected you will get a Connected status with a gold star as shown in Figure 3.15. You can now click the close box (the red X in the upper-right corner) and start using your wireless connection to access anything on the Internet your little heart desires.

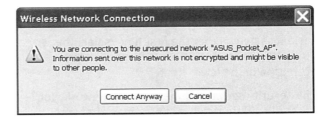

FIGURE 3.14

Windows will ask you for confirmation when connecting to an unsecured wireless network.

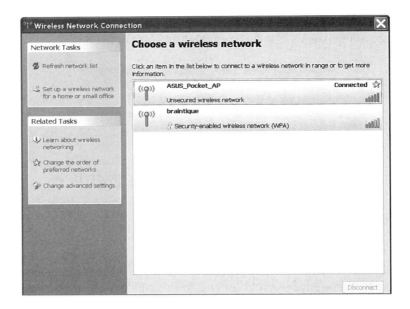

FIGURE 3.15

Successful connections to wireless networks are indicated by the Connected status and a gold star.

The process of connecting to a wireless network is something you will do often as you travel around with your mobile computer. I showed you how to achieve a connection starting at the Control Panel item called Network Connections because that let you see the wireless networking dialogs with all possible options.

However, Windows provides a more streamlined way to choose a wireless network. The system tray (in the lower right of your screen) has an icon for your wireless connection. Double-click this icon and choose View Available Wireless Networks from the context-sensitive menu as shown in Figure 3.16. Your computer will quickly find all the available wireless networks.

FIGURE 3.16

The quick way to find wireless networks is with the wireless connection icon in the system tray.

Advanced Wireless Network Security Settings

When you successfully connect to a wireless network for the first time, Windows auto-matically creates a connection profile for that wireless network. In most cases you will never need to access the profile. However, from time to time, there might be situations in which altering a profile is the only way you will get connected or respond to a change in the way the network is run (changing from unsecured to secured, for example). So, now that you're an expert at connecting, graduate to the advanced class where you will learn how to modify wireless connection profiles.

To set wireless network security settings

1. Click the Start button, and then select Control Panel.

2. Choose Network Connections in the Control Panel.

3. Select your Wireless Network Connection from the list on the right of the Network Connections window.

4. Click Change Settings of the Connection at the left of the window.

5. Select the Wireless Networks tab.

    The Preferred Networks section of the Wireless Network Connection Properties dia-log should now be visable (see Figure 3.17). There will be at least one entry with the same name as the wireless network you connected to previously.

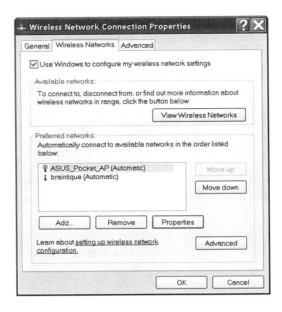

FIGURE 3.17

Automatically generated connection profiles are listed in the Preferred Networks section.

Click on a connection profile, and then click the Properties button to get access to advanced settings where you can change security-related options.

The Association tab (and the Authentication tab you can see behind it in Figure 3.18) is used to set encryption keys for the wireless network. Although entering this information can get complicated, it doesn't need to be.

An encryption key is simply a key, or formula, used to encrypt, or encode, information so that it is not easily readable. Generally, you need an encryption key twice—once to encode the information, and then once again to decode it.

First of all, if you are connecting to a public hotspot, it will almost certainly be running without encryption. In this case, the Data Encryption drop-down list should be set to disabled (and you probably don't need to make any other settings in this window).

Industry statistics show that as much as 90% of all private wireless networks are running without encryption. This is a bad practice that makes the wireless network vulnerable to infiltration as I explain in Chapter 18. If you are doing so, I hope that you'll switch to encryption after reading this book. But for the moment, if you are connecting to a private network that is running without encryption, once again all you need to do is to set the Data Encryption drop-down list to Disabled.

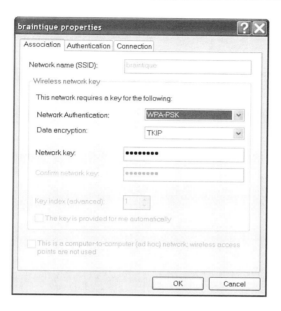

FIGURE 3.18

The Association tab of the Wireless Network Properties window is used to set encryption keys.

It's possible to provide encryption keys automatically, but this requires a special server, so most likely it will only be done in situations such as a corporate campus that is connected via wireless. If you are connecting to a wireless network that automatically provides encryption keys, check the This Key Is Provided for Me Automatically check box, as shown in Figure 3.19.

I'm leaving the most messy possibility for last. That is, the network is encrypted (as all private wireless networks should be), but the key is not automatically supplied. In this case, as you would suspect, to connect to the network, you need to know the key.

In this situation, documentation often says something such as "Contact your network or system administrator for information about your encryption key."

If you are like me, and have a small home or business network, at this point you throw up your hands and say, "But I am the bloody system or network administrator. Or at least, there's no one else here to do it if I don't." So where do you find the encryption key?

FIGURE 3.19

Check This Key Is Provided for Me Automatically if you are connecting to a wireless corporate network that provides this feature.

The good news is that you probably set the wireless network up. This means that you entered the encryption key in the first place, using mechanisms that I explain in Chapter 13, "Setting up Your Access Point," Chapter 14, "Configuring Your Wi-Fi Network," and Chapter 15, "Advanced Access Point Configuration." In those chapters, I advise you to make note of crucial settings, such as the encryption key. But even if you didn't write it down, you can open the access point's configuration utility (as explained in Chapter 13), and find out what it is.

To summarize, I've explained three encryption key possibilities:

- There is no encryption because you are connecting to a public network or a sloppy private network.

- A key is provided automatically by a special server, generally only the case in large corporate wireless networks.

- You know the encryption key and have entered it manually.

Whichever is the case, click OK to accept your encryption settings. The next step is to test your network.

Testing the Network

There's no magic to testing your wireless network. Essentially, there are three steps you can take:

- Inspecting the connection information provided by Windows XP

- Making sure that you can browse your entire network

- Seeing if you can connect to the Internet (provided your network has working Internet access)

As soon as you are successfully connected to a wireless connection, information about the successful connection will appear in balloon form above the network icon in the Windows taskbar, as shown in Figure 3.20.

FIGURE 3.20

When you are successfully connected, a balloon above the network icon in the Windows taskbar will provide connection information.

By the way, to see status information at any time similar to that shown in Figure 3.20, just hover the mouse above the wireless connection icon in the taskbar.

Returning for a moment to the Network Connections window, if you look at Figure 3.21, you can now see that the big red X, shown back in Figure 3.9, is no longer associated with the wireless adapter. This indicates that the wireless connection is working.

The context menu associated with the wireless connection (shown in Figure 3.22) provides a way to enable/disable the wireless connection, check its status, and perform other important operations.

TIP

You might also find it useful to know that when you are connected to a wireless network you can double-click the Windows taskbar icon representing the wireless connection, and the Wireless Network Connection Status window will open. Also, many options related to the wireless connection can be access by right-clicking the taskbar icon that represents it.

FIGURE 3.21

There's no longer a red X associated with the wireless adapter, which signifies that the wireless connection is working.

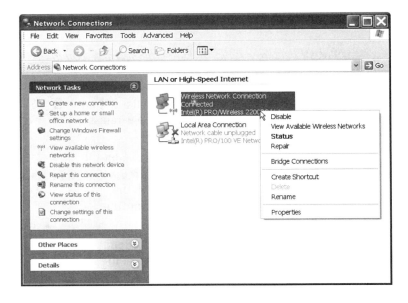

FIGURE 3.22

You can perform many tasks using the context menu for the wireless connection.

To see that your wireless connection is actually working, the best thing you can do is open an Internet browser and attempt to access something on the Internet. (This assumes, of course, that your network itself is connected to a router or access point/router combination with Internet access as explained in Part IV.)

Another nitty-gritty test to make sure wireless networking functions properly is to use Windows Explorer to connect to the other computers on your network (see Figure 3.23).

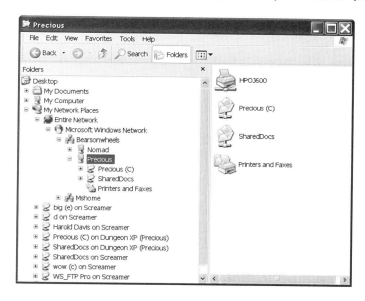

FIGURE 3.23

A good way to determine that wireless networking is working is to check that you can see the other computers on your network.

One final window you should know about is the Wireless Network Connection Status window, shown in Figure 3.24.

There are a number of ways you can open the Wireless Network Connection Status window, including

- Double-clicking the Windows taskbar icon representing the wireless connection

- Selecting the wireless connection in the Network Connections window and choosing View Status of the Connection on the Network Tasks pane

- Selecting the wireless connection in the Network Connections window, right-clicking, and choosing Status from the context menu

FIGURE 3.24

The Wireless Network Connection Status window shows the status of a wireless connection.

With the Wireless Network Connection Status window open, you can get information about connection status, network name, duration, speed, and signal strength. Signal strength is represented by a series of colored bars. (Five green bars is best; fewer bars indicate a weaker signal.) Chapter 8 provides some information about speed, so you can get a sense of whether your wireless connection speed is good enough for your uses. (You can never have a connection that is too fast.)

Clicking the Disable button on the Wireless Network Connection Status window disables a wireless connection.

Clicking the Properties button on the Wireless Network Connection Status window opens the Wireless Networks tab of the Wireless Network Connection Properties window (see "Using Wireless Network Connection Properties" earlier in this chapter).

Clicking the View Wireless Networks button on the Wireless Network Connection Status window opens the dialog that shows all the wireless networks within range.

Connecting to a Wireless Network Automatically

Let's go back for a moment to the Wireless Networks tab of the Wireless Network Connection Properties window. You might have noticed the Preferred Networks box at the bottom of the tab and wondered what this area is for (see Figure 3.25).

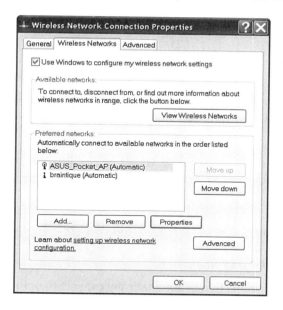

FIGURE 3.25

The Preferred Networks box is used to connect automatically to wireless "stations."

The answer is that the Preferred Networks box is used to connect automatically to given wireless signals. The entries in this drop-down list are automatically connected when your computer is turned on, which is convenient for home and office networks. (Who wants to have to manually connect each time?)

Furthermore, using the Move Up and Move Down buttons, you can determine the order of connection.

To add a network to the Preferred Networks box, click the Add button and enter its SSID. You'll also have to supply the encryption key if it is needed, as I explained earlier in this chapter.

As you might suspect, selecting a wireless network and clicking Remove removes the wireless network from the list of automatic connections.

TIP
The Properties button also lets you supply encryption information for a network listed in the Preferred Networks box.

Advanced Preferred Network Features

If you click the Advanced button on the bottom right of the Wireless Network tab, the Advanced dialog will open. The Advanced dialog, shown in Figure 3.26, allows you to set a number of options related to automatic connection.

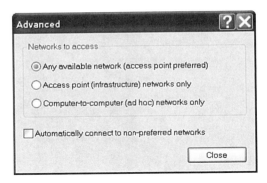

FIGURE 3.26

The Advanced dialog is used to set a number of properties relating to automatic connection.

Using the Advanced dialog, you can

- Set the wireless connection to automatically connect to both access points and ad hoc networks (access point preferred)
- Set it to connect to access points only
- Set it to connect to ad hoc networks only

The difference between an access point network and an ad hoc network is that an access point network broadcasts its signal from a central base station, whereas an ad hoc network involves peer-to-peer communication between computers. You'll find more about the distinction, and how to work with both kinds of wireless networks, in Part IV.

In the Advanced dialog, you can also check Automatically Connect to Non-preferred Networks. (This check box is shown at the bottom of Figure 3.26.) If this option is checked, the wireless connection will automatically connect to any available signal it can, even if no wireless networks are listed in the Preferred Networks box.

Enabling a Personal Firewall

A *firewall* is a blocking mechanism that helps to keep intruders out of a network (see Chapter 18 for more information about firewalls).

If you are connecting your Centrino laptop to a private network that is itself connected to the Internet, you are probably already protected by a firewall built in to the network hardware (the router) and do not need to enable a personal firewall.

However, if you are connecting to the Internet directly with your Centrino laptop, for example at a public hotspot, it is a wise precaution to enable the personal firewall software built in to Windows XP. (See Chapter 17, "Protecting Your Mobile Computer," for more information.)

To enable the Windows XP personal firewall, open the Windows Firewall item in the Control Panel. Be sure the On radio button is selected. Finally, click OK to preserve the setting.

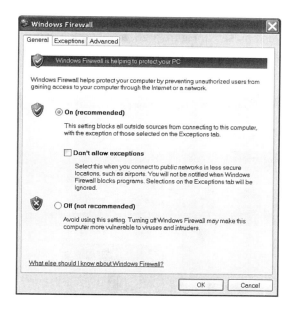

FIGURE 3.27

Enabling the personal firewall software built in to Windows XP helps protect your computer when you are directly connected to the Internet at a public hotspot.

Bridging Wireless Connections

A network bridge provides a way to connect different segments of a local area network (LAN). Networking bridging can be achieved either using hardware or just using software. It is explained in more detail in Part IV.

NOTE

With Windows XP Service Pack 2 (or "SP2" as it is affectionately known), the Windows XP personal firewall is enabled by default. In other words, it will be turned on unless you turn it off. So if you've installed SP2, you don't need to take the steps outlined here to enable the Windows XP personal firewall.

It's easy to use Windows XP to create network bridges out of the connections on your mobile computer. For example, your computer might have both a wired network connection and a wireless network connection, or it might have a connection to two wireless networks. If both the wired and wireless network connections were placed in the bridge, or if the two wireless connections were placed in a bridge, effectively a new larger network has been created, using the pairing in your mobile computer as a through point.

To add two (or more) connections to a wireless bridge, select both connections in the Network Connections window. Right-click, and choose Bridge Connections from the context menu. As you can see in Figure 3.28, a Network Bridge will be added to the Network Connections window, and the bridged connections will now appear as part of it.

FIGURE 3.28

You can use the Network Connections window to bridge connections.

If you need to reverse the process, it's easy to remove a wired or wireless connection from a network bridge by selecting one or more connections in the Network Connections window, right-clicking, and choosing Remove from Bridge from the context menu.

Summary

Here are the key points to remember from this chapter:

- The wireless network configuration software built in to Windows XP might seem daunting, but it is really pretty simple to use.

- There are many ways to get to the same wireless configuration windows.

- No matter what wireless mobile laptop you have, Windows XP wireless configuration works in the same way.

- To connect to a wireless network, you have to configure your computer with a name, and as part of a workgroup, as you would with a normal wired network.

- The Network Connections window is organized so that you can perform the key tasks related to setting up networks.

- The Network Tasks pane of the Network Connections window provides wizards (and other tools) for helping you with configuration tasks.

- The Wireless Network Connection Properties window is the central command station for connecting to wireless networks.

- If a wireless network is encrypted, and it doesn't automatically provide a key, you'll need to know the encryption key.

- You should test your wireless connections by connecting to the Internet and your network.

- It's easy to set your wireless connections for automatic connection—for example, to your home or small office network.

- Many settings are possible related to wireless networking; the most important are explained in this chapter, along with references to the places in this book you can learn more about them.

- Don't worry, be happy! If you follow the directions in this chapter, you'll have no problems connecting to a wireless network.

GETTING THE MOST FROM YOUR MOBILE COMPUTER

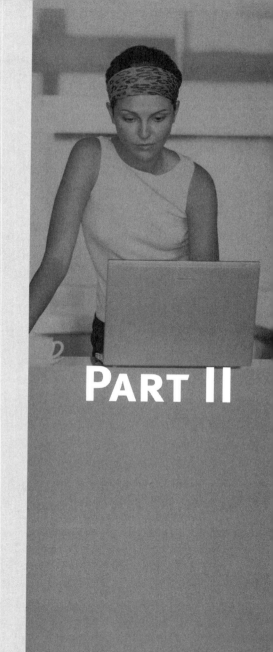

PART II

Software That Makes the Most of Mobile Computing

4

I f you are a mobile computer user, you are probably on the go. You need to be able to use your computer anywhere—from home, or home office, or on the road. It needs to work equally well from all places, anytime, anywhere!

If you are this kind of user, your laptop that uses Intel Centrino mobile technology *is* the brains of the operation. (Well, maybe you add a little something, too!)

So, if you are the on-the-go kind of mobile user, this chapter is for you. I'll show you how to make the most out of the software that is available for you and how to solve common on-the-go business challenges. You'll learn how to work equally well from your home or home office and anywhere there is connectivity, thanks to the nifty features of your mobile computer, and the wonders of wireless networking.

Office Suites

Probably the most important piece of software on your computer is a so-called Office Suite. This set of programs typically includes word processing capabilities, the ability to create and manipulate spreadsheets, and much more.

With an Office Suite installed on your mobile computer, you can do all the work that any office needs, from anywhere.

As you probably know, the world's most popular office suite is Microsoft Office. Microsoft Office comes in many versions and flavors, with different programs included depending on the version, and with widely varying prices.

Perhaps a version of Microsoft Office was pre-installed when you bought your laptop. If so, this was likely the Microsoft Office Small Business Edition, which can be purchased (if you don't already have it installed) for a retail price of $449.

The Microsoft Office Small Business Edition 2003 includes the following programs:

- Microsoft Word (for word processing)
- Microsoft Excel (for working with spreadsheets)
- Microsoft Outlook (email and personal information management)
- PowerPoint, a program used to create presentations (see "Powerful Presentations on the Go," later in this chapter)
- Microsoft Publisher (an easy-to-use desktop publishing program)

For a little more money than the Microsoft Office Small Business Edition 2003, you can have the Microsoft Office Professional Edition 2003. The Professional Edition adds to the suite

- Access, an end-user database program

Although you almost certainly know about the Microsoft Office suites, you might not know that there are some viable alternatives to Microsoft Office—which are much less expensive, or even free. For the most part, alternative office suite products are file compatible with Microsoft Office, meaning that files created in one of these programs can be opened in Microsoft Office, and vice versa.

Alternative office suites available for Windows include

- Corel WordPerfect Office suite, which includes the WordPerfect word processing program, Quattro Pro for working with spreadsheets, and a presentation tool similar to PowerPoint, available for about $80

- OpenOffice, which is available as a free download from `http://www.openoffice.org`, and includes a word processor, a spreadsheet application, a presentation tool, and more

- Sun Microsystems's StarOffice suite provides a word processor, a spreadsheet application, a presentation tool, a graphics tool, and a database application, all for a suggested retail price of $80

Powerful Presentations on the Go

For most business users of mobile computers, the ability to create and share presentations is of the utmost importance. This is because often the whole reason for undertaking business travel is to make a presentation—for example, in order to sell a product or service or to explain a business process or technology.

Generally, PowerPoint is the software of choice for creating these presentations, also called slide shows, or *decks*.

As I mentioned earlier, you might already have PowerPoint as part of the Microsoft Office Professional Edition. If not, you can buy a copy of Microsoft PowerPoint 2003 for a list price of $229. (Yes, because this single program costs about as much as the Microsoft Office Small Business Edition suite when bundled with a computer, you might just as well buy the whole thing.)

I have given hundreds of PowerPoint slide shows for a wide variety of purposes. When PowerPoint works well, and you've prepared a good deck, it is a thing of wonder and grace. However, a PowerPoint presentation given at a place you've never been to before is Murphy's law waiting to happen—a lot can go wrong. A PowerPoint slide show that is full of fancy effects such as fades and dissolves, or in which the speaker simply reads the slides, can be dreadful—and self-defeating.

I can't tell you everything about creating slide shows with your mobile computer without writing a whole book just on the topic. But if you follow these simple guidelines, your presentations are likely to be effective and trouble free.

Creating the Slide Show

It's really important to keep your audience in mind when you create your slide show using PowerPoint (or another presentation tool). You need to understand who you are "speaking" with, and also to respect your audience. Don't bore them!

Creating a slide shows involves writing, visuals, and design—so it really can be a pretty complex undertaking. That said, it is pretty much true that (when it comes to a presentation) simpler is better. The following guidelines should help:

- Your slide show is not your speech: Create a slide show that consists of talking points for what you are going to say, but don't just read the slides. Verbatim reading of a slide show is boring, boring, boring and bad, bad, bad.

- Keep the big picture in mind when you create your presentation. Know what you want to say before you start trying to say it. A PowerPoint presentation is essentially a white paper in outline form. Each slide should be a topic.

- Use a PowerPoint template when you create your presentation. A simpler one is best.

- Creating a personalized template is a great way to make your presentations more professional. You can easily create a personalized template by modifying a few elements in one of the templates that ships with PowerPoint and by adding your company name or logo.

- Be careful to choose background and text colors that make your slides readable at a distance.

- Font size needs to be big enough so that even those in the back of an audience can read the slides. In no event should you use a font size less than 14 points. While we're on the topic of fonts, choose a simple font. (Arial or Times New Roman are fine.) Don't mix and match font families (in no event use more than two font families, and one is better).

- Better presentations are svelte: I have actually sat through PowerPoint decks with hundreds of slides, and I can assure you that shorter is better, and far more effective.

- Always leave room for questions and dialogue with your audience.

- Break up your words with visuals such as flow charts and pictures. Visio is a great and easy-to-use tool for creating flow charts and system diagrams.

- Once and for all, eschew those fancy effects. I've already mentioned that you don't need fades and dissolves between slides. (They alienate the audience, and make them giggle.) You also really, really don't need to individually add elements to each slide (where you click the Next button, and elements such as text loop one by one onto the slide). This wastes time and drives your audience crazy (not in a good way).

Presentation Mechanics

If a tree falls in a forest, and no one sees it fall, has it really fallen? This pseudo-Zen question is by way of driving home the point that all the work you spent on creating your presentation will be for naught if you have no way to show it.

Following these guidelines will help minimize the chances of getting caught in a show stopper:

- If possible, test drive your slide show with the projector you will be using, in the location for the presentation. If you can't get to the location, check to see what kind of projector will be used and see if you can test with a projector of the same type locally.

- Check in advance to see if you will have wireless connectivity at the site of the presentation. If you do, you'll be able to connect to your home network in an emergency—for example, you lose the only copy of your presentation—and use some of the other file sharing options explained later in this chapter.

- Bring redundant copies of your presentation.

- Make sure that you are showing the final version. (You can name it using the word "final.")

- Use the PowerPoint Pack and Go Wizard to create an independent version of your presentation, complete with viewer. Burn the slide shown onto a CD-ROM, and take that with you (in addition to your mobile computer) as one more form of redundant backup.

Email

As you probably know, it is easy to use Microsoft Outlook as a program, sometimes called an *email client*, from your home or office to download, compose, and respond to email. Outlook ships as part of the Microsoft Office suite described earlier in this chapter. A stripped-down version, Outlook Express, is available as a free download, if you don't have it already. (You'll usually find Outlook Express pre-installed as part of Microsoft Windows.)

There are also some other email client programs that proponents swear by. For example, you can find out more about Qualcomm's Eudora Email at `http://www.eudora.com`.

From the viewpoint of the mobile computer user, there aren't any major issues about using an email client when you are on your home or small office network. But managing email on the go can be quite a challenge.

One problem is that you might not be able to access your email server with your email client (for example, Outlook) when you aren't connecting from your home (or small office). This depends on your ISP. But many broadband vendors set up their servers (for security reasons) so that you can only use a regular email client from home.

If your ISP is set up this way, hopefully they have provided a way for mobile computer users on the go to get around this problem. A common solution is for the ISP to provide Web-based email as well as standard email. When you are on the road, you simply use the Web interface to read, compose, and send your email.

TIP Remember that if you are using Web-based email from a public hotspot, it is not secure.

Figure 4.1 shows the Comcast Web-based email interface.

So the great news about a Web mail interface is that you can use it from anywhere, as long as you are connected.

What if your ISP does not provide a Web mail alternative? There are still several viable alternatives.

One is to get a free email account just for use while you are on the go. Hotmail, `http://www.hotmail.com`, and Yahoo!, `http://mail.yahoo.com`, are two of the most popular providers of free email, which use a Web-based interface. Google recently entered the fray with Gmail, `http://mail.google.com`.

NOTE Email accounts, such as those provided by Hotmail, Yahoo!, and Google are free, but they usually do expire if you don't use them—so be careful if you think of one of these accounts as your permanent email address. There is also typically a line of advertising at the bottom of each email you send.

Free Web-based email works well for sending email on the go, but it doesn't provide a mechanism for picking up email from your regular account. You can give out the free email address as your primary email contact (and forget about the email service your ISP provides), but this has some drawbacks:

- You don't get to use Outlook (or some other powerful email client) when you are "at home"

- Free email accounts do expire, so you risk "losing" an email address given out as professional contact information

- Some people feel that free email accounts do not project an entirely professional image (but I say, "what is wrong with free?")

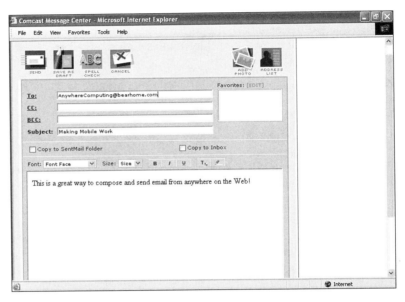

FIGURE 4.1

If your ISP provides a Web mail interface, you can use it from anywhere on the Internet.

An attractive alternative for a small business professional is to set up an email forwarding arrangement using a service such as that provided by Domain Direct, http://www.domaindirect.com. With Domain Direct, you register a domain name—for example, www.yourname.com. This costs $25 per year or less (depending on the number of years you purchase) .

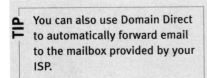

TIP
You can also use Domain Direct to automatically forward email to the mailbox provided by your ISP.

You then use the Domain Direct control panel to set up an email address or addresses—for example, jane@yourname.com and tarzan@yourname.com. When you are at home, your email program can be set to pick up email from the domain you set up with Domain Direct. When you are mobile, you can use Domain Direct's Web mail facility.

Auto-Responding

Another issue that confronts mobile computer users when it comes to email is to auto-respond or not to auto-respond. I'm sure that it is nobler to auto-respond, but there are some reasons you might not want to.

An auto-response is a message automatically generated and sent in response to an incoming email message. Typically, it says something similar to "I'm away at the Grand Kauai Shores Hotel until the 22nd, and I'll reply to your message then."

You can use an email program such as Outlook to generate auto-responses, but this isn't a very good idea. (For one thing, it means that your computer and email program need to stay connected to the Internet for the entire time you are gone.)

It works much better to use your Web mail interface to set up an auto-response. Both a Web mail interface provided by your broadband ISP and a third-party provider (such as Domain Direct) will provide this facility.

Here's why you might want to provide an auto-response while you are away on the road:

- It's rude and unprofessional to leave email unanswered.

Here's why you might want to think twice about using auto-responses:

- Auto-responses verify to spammers that your email address is "real."

- If you work out of a home office, or a small office, auto-responses give anyone who has your email address the idea that you are not physically there. (Would you leave a note on the front door of your house?)

Like many things in life, the choice is yours. But if you are mobile computing and plan to check email anyway, why not simply respond to it, and forego the auto-response?

Web Browsers

I am sure that you are familiar with Microsoft Internet Explorer as a Web browser you can use to surf the Internet. You might not know that a number of other Web browsers are available that—in the opinion of many professionals—provide better security; features such as pop-up, spyware, and virus blocking not available in Explorer; and faster page downloading.

Of course, you'll have to decide for yourself which Web browser you prefer using when you are on the go. But why not at least try some of the alternatives before you settle on Explorer as the only choice?

Alternative browsers include

- Firefox, available as a free download from the Mozilla project,
  `http://www.mozilla.org`

- Navigator, available as a free download from Netscape, `http://www.netscape.com.`

- Opera, available as a free download with banner advertising or in a paid version
  without ads for $39 from Opera, `http://www.opera.com.`

Should you decide to use Internet Explorer, when you prepare to take your mobile computer on the road, particularly if you will be using it at public hotspots, you should pay attention to some of the settings in the Internet Options window.

To open the Internet Options window, choose Internet Options from Internet Explorer's Tools menu. You should probably spend some time looking through the various tabs and settings in this window, but—to start with—you should make sure that Security and Privacy are set high enough.

To set Security, open the Security tab, shown in Figure 4.2.

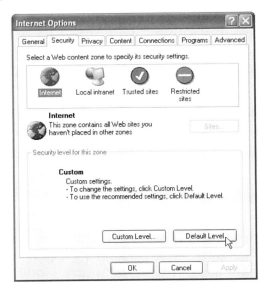

FIGURE 4.2

Use the Internet Explorer Security tab to change security options.

In the Security tab, at a minimum, you should select the Internet icon shown on the left of Figure 4.2, and click the Default Level button to make sure that default levels of security are in place.

Next, click Privacy to open the Privacy tab, shown in Figure 4.3.

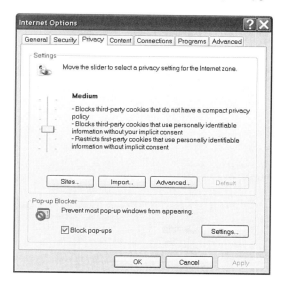

FIGURE 4.3

Use the Internet Explorer Privacy tab to change privacy options.

The slider, shown on the left-hand side of Figure 4.3, is used for privacy settings. At a minimum, privacy should be set to Medium, as shown in Figure 4.3.

I don't leave home and go mobile without a really cool add-on to Internet Explorer, the Google Toolbar, which is shown in Figure 4.4.

The Google Toolbar can be downloaded free from `http://toolbar.google.com`. As you'd suspect, it offers you the ability to search Google no matter what web page is open in your browser. It also offers some really useful features, such as a pop-up blocker and the ability to auto-fill Web forms with your personal information (so you don't have to re-type it all the time).

> **TIP** You can tweak the security settings by clicking the Custom Level button.

> **TIP** High privacy, rather than Medium Privacy, is even more appropriate when using a public hotspot.

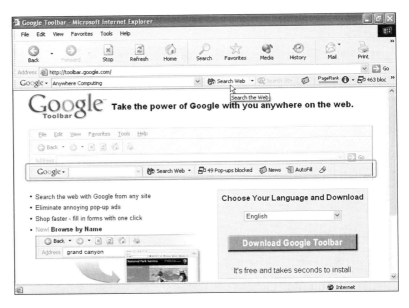

FIGURE 4.4

The Google Toolbar is a great Internet Explorer add-on for mobile users.

Transferring and Synchronizing Files

You need to have the right files when you are on the go. There are two problems related to having the right files that mobile users need to deal with:

- Getting the files you need from a remote location

- Making sure that files you've worked on while you were traveling match the ones on your home computer

Fortunately, there are reasonably good ways for today's mobile computer user to deal with both these issues.

Getting Files

Let's say that you're far away from home and about to give a presentation. You boot up your mobile computer, and—oops—notice that your PowerPoint is missing. Or maybe you notice that you forgot the most recent version at home. (As you might have noticed, people are always tweaking their PowerPoint decks.)

No problem! You connect to the Internet using the wireless network provided by the conference center—or perhaps by your hosts—and download the file from your network.

To facilitate this scenario, you should have a virtual private network (VPN) set up, as I explain in Part V, "Securing Your Computer and Network."

Both small and large businesses that expect to have remote users download a great many files often set up a FTP—File Transfer Protocol—server to facilitate this.

You can connect to an FTP server using Internet Explorer, or even your computer's command line, but it is much easier to use a dedicated FTP client program if you expect to be doing a great deal of file transfers this way.

A quick search using Google will show you that a great many FTP client programs are available for download. Many of these are free or available as shareware. All are inexpensive.

One of the best is WS_FTP Pro, available from Ipswitch, `http://www.ipswitch.com`. Because Ipswitch makes available a free 30-day trial of the software, you can try it for yourself and see if you like it.

Figure 4.5 shows a portion of the WS_FTP interface.

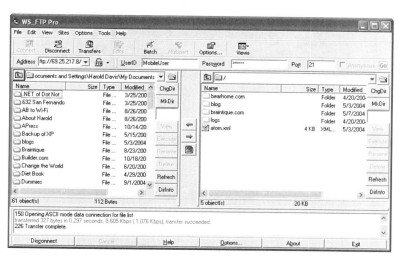

FIGURE 4.5

WS_FTP Pro is one of the best FTP client programs.

If you look at Figure 4.5, you'll see two panes. The pane on the left shows the folders and files in your mobile computer. The pane on the right shows the FTP server on your home network. You can transfer files by dragging and dropping from one pane to the other.

Aren't ready to set up a VPN or an FTP server on your home network? Don't despair: there are other easy ways to make sure that your files and data are available remotely wherever you have an Internet connection, when you need them.

Many online services offer the ability to store your files—or even the entire contents of a computer or computers—on the Internet. Depending on the amount of storage you need, these services range from free to fairly inexpensive. Not only can you use services such as these to retrieve files when you need them, where you need them, but this will also work for you as an "offsite" backup—provided (as with any back up software) you are prepared to regularly back your computers up.

For example, Xdrive, at `http://www.xdrive.com`, will store 5 gigabytes worth of files and data for $10 per month. (You can get a free trial.)

To make the offsite storage easier, Xdrive provides a desktop utility, shown in Figure 4.6, that allows you to drag and drop multiple files to the Xdrive facility.

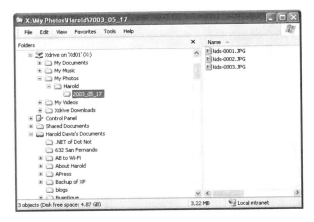

FIGURE 4.6

The Xdrive desktop program integrates with Windows Explorer.

The program shown in Figure 4.6 integrates with Windows Explorer. The Xdrive facility is designated with an X. You simply drag and drop files and folders to the X-marked area the way you normally would in Windows Explorer.

When you are traveling with your mobile computer, and you realize that you need to retrieve some of your files from Xdrive, you can simply connect to the Internet using a hotspot or other wireless network, and use the Web browser–based interface shown in Figure 4.7 to download the files you need.

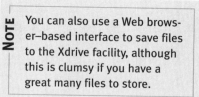

NOTE You can also use a Web browser–based interface to save files to the Xdrive facility, although this is clumsy if you have a great many files to store.

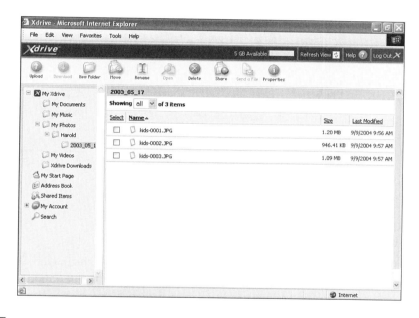

FIGURE 4.7

You can retrieve files from Xdrive using a standard browser window.

Synchronizing Files

Here's the scenario: You leave home (or office) with your mobile computer, work on some documents, or presentations, that you began earlier at home. When you return home, you want to make sure that you have the most recent "worked on" versions on your home (as well as mobile) computer.

Windows provides an easy mechanism for dealing with this scenario, called a *Briefcase*.

To create a Briefcase, in Windows Explorer, select New from the File menu, and then choose Briefcase as shown in Figure 4.8.

| Chapter 4 | Software That Makes the Most of Mobile Computing |

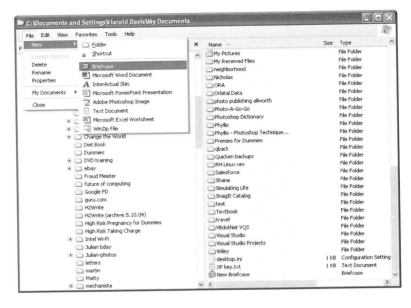

FIGURE 4.8

It's easy to create a new Briefcase in Windows Explorer.

The new Briefcase is shown in the lower right of Figure 4.8.

> **TIP**
> You can create a new Briefcase in the same ways you can create a new folder in Windows Explorer.

Once the new Briefcase has been created, you can work with it in the same way you work with folders, renaming it and dragging and dropping files into it.

When you are ready to go on the road, with your mobile computer connected to your wireless network, copy the Briefcase and its contents to your mobile computer.

Have fun! Have a great trip. Do a lot of work on the contents of the Briefcase.

> **TIP**
> You can synchronize only selected files in a briefcase, rather than all the files, by selecting the ones you want to bring into synch, and then choosing Update Selection from the Briefcase menu.

When you return, connect you mobile computer once more to your wireless network. Using Windows Explorer, open the Briefcase. On the Briefcase menu, choose Update All. The files in the Briefcase will be synchronized.

Pop-Up and Spyware Blockers

Pop-ups are browser windows that open while you are surfing the Internet, usually without your consent, usually advertising something. They are a nuisance. In contrast, spyware—programs placed on your computer without your informed consent that monitor your activities—can do actual harm (in addition to delivering information about your habits to third parties).

The general category of these kinds of software has been aptly called *malware*—because they range from the merely obnoxious to the downright malicious. Some species of malware cut across the categories: One nasty piece of work takes over the search features of Internet Explorer, reports on your activities, and opens pop-up ads targeted at you based on your searches and the reports.

> **NOTE** There's nothing to stop you from using a Briefcase to synchronize files remotely. For example, if you are using a VPN to connect to your home network, as I explain in Part V, you might want to synch the files in your Briefcase from time to time with the versions at home.

By the way, it's fine if you decide to install this kind of software by choice, but absolutely not in my book if you haven't agreed to it, or if the agreement is in some fine print that you've never read. An awful lot of this annoying software gets installed as a hidden "bonus" with some shareware and freeware programs.

As I mentioned earlier in this chapter, if you are using a browser that is an alternative to Internet Explorer—such as the Mozilla project's Foxfire—it might already have effective pop-up and spyware blocking. Any add-ons to Internet Explorer, such as the Google Toolbar, provide pop-up blocking (and even tell you how many it has blocked). In the due fullness of time, pop-up and spyware blockers will be incorporated into antivirus software and even into operating systems.

Protection is good, but it does involve trade-offs. Software that constantly monitors for spyware can slow your computer down, and depending on the your mobile connection speed, you might decide not to run this kind of software. Even so, don't forget the need for protection. You can run a program that scans your system for spyware every so often to make sure that you are "clean"—without negatively impacting ongoing performance.

There's really no downside to blocking pop-ups. This may be handled for you by your browser or browser add-on that blocks these obnoxious advertisements. If not, you can run dedicated blocking software on your mobile computer.

It's true that some sites and applications use pop-up windows for functional reasons, such as allowing a user to log in or to start a remote session. If you need access to a feature such as this, you can always manually unblock a specific site. (All the pop-up blockers have both manual overrides and mechanisms for unblocking specific Web addresses.)

> **TIP**
>
> The SP2 (Service Pack 2) release of Windows XP has improved security features, including its own pop-up blocker. In addition, the personal firewall program that ships with Windows XP now defaults to "on" (rather than "off" as it did prior to SP2).

If you do a search on Google for pop-up and spyware blockers, you'll find many offerings. Many effective products perform these services, so you have quite a number to choose from.

A word of warning here is to be careful about any pop-up and spyware blockers offered as part of spam emails. These "products" might install spyware themselves and should not be trusted. (In any case, the best way to eliminate spam is to never "click through" it; if no one ever bought anything because of spam, there would be no spam email.)

Spy Sweeper, available as a free download from http://www. webroot.com is a program that will reliably detect and eliminate any spyware installed on your system. Figure 4.9 shows the results of a Spy Sweeper system scan.

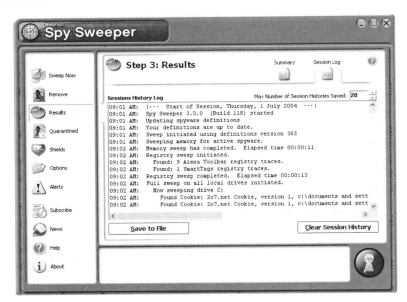

FIGURE 4.9

Spy Sweeper will scan your system for spyware and remove any malware it finds.

A good commercial anti-spyware program is also available from McAfee, `http://www.mcafee.com`. Although this program does have an initial cost (about $40), this is a reliable program (which also foils adware and obnoxious pop-ups when run in auto-protect mode) from a mainstream vendor.

> **NOTE**
> Spy Sweeper can be run to occasionally scan your system, or it can be used as an active shield against malware. If it is actively scanning, it will slow your system down. Note also that the initial Spy Sweeper program is free, but to keep it current you will need to purchase a subscription.

Summary

Here are the key points to remember from this chapter:

- All mobile users need an office suite of software, but there are some good (less expensive) alternatives to Microsoft Office if it didn't come pre-loaded when you bought your computer.

- If you follow some simple guidelines, you can make great slide shows for your presentations on the go.

- Mobile users need to consider their mobile email needs in advance.

- Consider alternatives to Internet Explorer, such as the Mozilla project's Firefox.

- It's easy to set things up so that you can download your personal files, anytime, from anywhere.

- It's a good idea to protect your computer from malware such as pop-up advertising and spyware.

Taking Digital Pictures from Your Laptop

It's very cool that many laptops with Intel Centrino mobile technology come with built-in digital cameras. This means that when you're out there, anywhere, with your mobile computer—and you've forgotten to bring along a digital camera—you can still take pictures. It's unclear how many Centrino mobile computers come with this feature, but probably at least 10 percent of new units purchased do. If your computer includes a camera, you can then transmit the pictures using Wi-Fi for processing or to a home network.

This chapter shows you how to take pictures with your Centrino laptop (provided it has a camera), how to get the most out of the camera, and some exciting things you can do with the pictures you've taken.

Picture Quality

A few years back, a friend of mine quipped that playing music on his computer was in essence turning an expensive computer into a lousy stereo. Well, times have certainly changed, and you can now play music with perfectly fine audio quality using your computer (or your iPod). But the camera in your Centrino laptop is likely to still be stuck in the place the sound card used to be—it's probably a low quality digital camera, comparable to the one in your cell phone. No doubt, in time, this will change for the better. I fully expect, in the next few years, to see digital cameras bundled in mobile computers that are every bit as good as standalone cameras (although this will take some form factor engineering).

DIGITAL CAMERAS IN MOBILE COMPUTERS AND CELL PHONES

As a matter of fact, the digital camera in your computer is likely to be quite similar to the one in your cell phone (and some other digital devices). In fact, it probably uses the same camera on a chip, which is likely made by a company called OmniVision Technologies (although there are many other manufacturers of cameras on a chip intended for cell phones and digital devices, ranging from tiny startups to giants such as Sony).

OmniVision is the world's leading supplier of single-chip camera solutions, which are baked into devices including mobile computers, cell phones, cars, toys, bar code readers, medical appliances, and more.

You can learn more about the OmniVision camera on a chip at `http://www.ovt.com`.

For right now, good enough will have to do, and it's fine for many uses. The typical digital camera built in to a mobile computer produces images with a maximum size of 640×480 pixels for a total final image size of about 300,000 pixels (or .3 megapixels). Obviously, this compares unfavorably with the current crop of consumer digital cameras, which routinely produce images from 2,000,000 to 4,000,000 pixels (2 to 4 megapixels) in size.

One of the key issues in taking pictures is being there—and having the camera—to take the picture in the first place. Obviously, a 640×480 pixel camera built in to a cell phone or mobile computer is no substitute for a higher quality dedicated digital camera. However, the whole point of having a camera built in to a mobile computer is that you might not have a separate camera available—and don't need to lug around two devices to take the occasional, perfectly adequate picture.

For example, suppose that you are sitting in your local coffee shop surfing the Internet with your cool, little Centrino laptop when two attractive members of the opposite sex come up to admire your hot mobile machine. This actually happened to me recently. If you have a camera built in to your Centrino laptop, you can offer to take their picture. It probably wouldn't have the same impact if the camera weren't built in to the computer.

A little more seriously, think of all the times you have your mobile computer with you, but not a camera, and have seen something that you'd like to photograph as a visual record and reminder.

WHAT CAN YOU DO TO IMPROVE THOSE IMAGES?

What can you do to tweak those images taken using the camera built in to a cell phone or mobile computer?

Actually, quite a lot! First, you should pay attention to the lighting conditions when you take the picture. Photos taken with these cameras work best when there is some contrast in light levels and colors, but not too much. They do not work well in low light conditions.

They are digital cameras; you are not wasting film when you take multiple pictures. Check to see how your photograph came out, and if it doesn't look good, see what you can do to fix the problem on the spot.

You can correct problems in your digital photos such as "red eye" by using photo retouching software such as Photoshop Elements, which has an easy-to-use tool for just this purpose. (The comparable mechanism in Photoshop CS is a preset for removing red eye, which is part of the Color Replacement Tool.)

Generally, the appearance of low resolution digital photos can easily be improved in photo manipulation programs as Photoshop (or Photoshop Elements) by making the image smaller and increasing sharpness and brightness.

Locating the Camera

If your mobile computer has a digital camera, the lens will most likely be located in the top center of the cover. This has the potential advantage of enabling the camera lens to "swivel" so that you can take pictures facing both "out" from the laptop and "in" toward the laptop's user.

For example, Figure 5.1 shows the Sony Visual Communication Camera VCC-U01, embedded in a Sony Vaio TR3A notebook with Intel Centrino mobile technology, facing outward.

When the camera component is swiveled inward, as shown in Figure 5.2, the camera lens faces the mobile computer's user.

Unlike some integrated digital cameras, the Sony VCC-U01 does not automatically focus. Focusing, or setting the distance the camera is from the desired subject, is accomplished using the dial you can see at the top of Figure 5.1 and 5.2.

Chapter 5	Taking Digital Pictures from Your Laptop

FIGURE 5.1

The camera integrated with this Sony notebook can face inward or outward. (Outward is shown.)

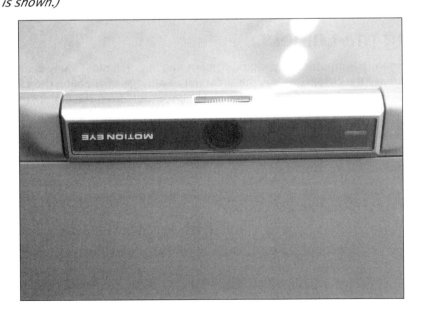

FIGURE 5.2

When the camera assembly is swiveled inward, the lens faces the mobile user.

Taking a Picture

The most straightforward way to initiate taking a picture on the Sony Vaio is to press the Capture button, shown on the right side of Figure 5.3. (The software on your mobile computer equipped with a digital camera might work a little differently than the Sony's, but it will probably follow the same general pattern.)

FIGURE 5.3

To take a picture, press the Capture button (shown on the right side).

After you press the Capture button, the Network Smart Capture software, shown in Figure 5.4, will open.

Use the dial that is part of the camera assembly to focus your picture, and click Capture. The picture will be added to the gallery—a kind of vertical film strip of photos—shown along the right side of Figures 5.4 and 5.5.

To view a picture you have captured (and that appears in the gallery), select it in the gallery portion of the Network Smart Capture interface (as shown in Figure 5.5), and click the Display Image icon on the upper right of the application. (The mouse cursor points to the icon in Figure 5.5.)

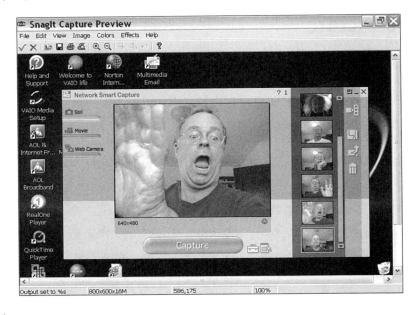

FIGURE 5.4

After you've pressed the hardware Capture button, the Network Smart Capture software will open.

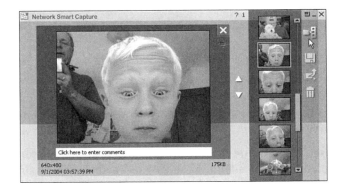

FIGURE 5.5

You can display captured images by toggling the Display Image icon.

To return to the capture window, toggle the Display Image icon again.

You might have noticed that it is possible to use the Network Smart Capture application to take movies and as a webcam, in addition to its use for capturing still digital photos. (Click the icons shown in the upper left of Figure 5.4 to start using the software and camera in these modes.)

By the way, it is also possible to open the Network Smart Capture application from within Windows XP (without using the Capture button). If you have a Sony notebook equipped with a digital camera, you'll find a Network Smart Capture program group when you click All Programs from the Windows XP Start menu.

I'll be explaining some of the camera configuration options available with the Sony camera and the Network Smart Capture software in the next section of this chapter.

Using the Microsoft Scanner and Camera Wizard

You should also know that you can use the camera built in to the Sony with some alternative software applications. For one thing, the Microsoft Scanner and Camera Wizard, found in the Accessories group of Windows XP Program Files on the Start menu, will walk you through the process of taking and saving pictures.

This wizard can be found on all computers running Windows XP, not just the Sony. It does lack any capability to control and configure the camera. So the trade-off is that any mobile computer will have the Scanner and Camera Wizard—but it won't give you much flexibility in working with the digital camera built into your laptop.

Figure 5.6 shows the panel of the wizard used for capturing photographs. When you click the Take Picture button, pointed to by the mouse cursor in Figure 5.6, the previewed image gets added to the picture library (shown on the right side). Note that you can rotate the previewed picture, using the icons to the right of the Take Picture button shown in Figure 5.6, but there are no other camera settings that you can configure.

Working with Your Camera

Returning to the Network Smart Capture software—and, by the way, you'll find comparable software on any mobile laptop with a digital camera—there are plenty of ways to tweak your photographs (and have fun while you are doing so).

To get started, click the Change Settings icon, found to the right of the Capture button shown in Figure 5.4. The Still Image Setting window will open.

Have a look at the Capturing Format tab, shown in Figure 5.7.

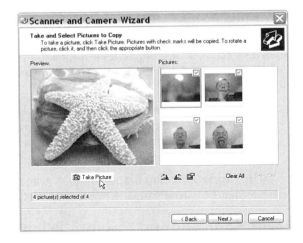

FIGURE 5.6

Click the Take Picture to add an image previewed in the camera to the Picture library.

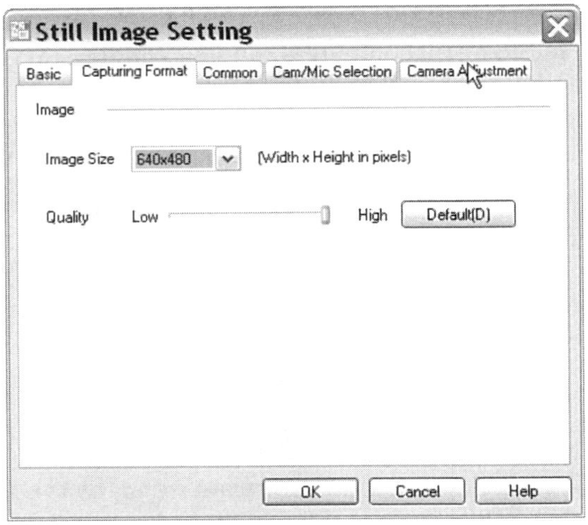

FIGURE 5.7

The Capturing Format tab allows you to set camera resolution.

Using this tab, as shown in Figure 5.7, you can set camera resolution up to its maximum.

If you click on the Camera Adjustment tab, shown in Figure 5.8, you'll see that you can adjust a great many camera settings using sliders, including Brightness, Color, Contrast, Hue, and Sharpness.

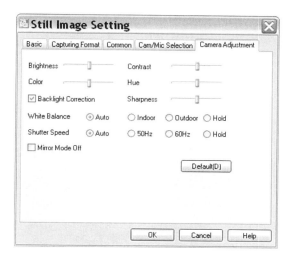

FIGURE 5.8

The Camera Adjustment tab can be used to control Brightness, Color, Contrast, Hue, Sharpness, and more.

Close the Still Image Setting window. Click the Select Capture Menu icon (found to the right of the Change Settings icon, and shown in Figure 5.4). A palette of special effects will open (see Figure 5.9).

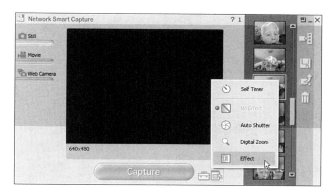

FIGURE 5.9

You can choose from among a number of different effects.

These effects include the ability to zoom, to set a self-timer, and to include photo manipulations, such as vignetting. For example, you could choose Spotlight from the Special Effect library window shown in Figure 5.10.

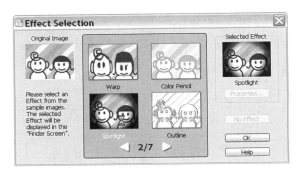

FIGURE 5.10

The Special Effect library window allows you to choose an effect to be applied.

With Spotlight selected as an effect, a spotlight is applied to the images you take with the camera, as you can see in Figure 5.11.

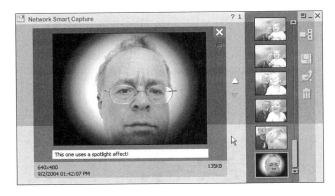

FIGURE 5.11

The spotlight effect has been applied to the image taken by the camera.

Saving the Photos

Taking pictures used to be a cumbersome affair. Well, not quite as bad as the nineteenth-century plate cameras of back breaking weight with black tents for loading film plates. But still, you had to load film in to the camera, wind it, rewind it, remember not to open the camera until the film had been rewound, and get the film to a photo lab. Some time later—it might be hours, and it might be days—your photographs would be ready. If you wanted to share them with a friend, you had to go back to the photo lab and have new prints made.

The best picture is the one you take. If you have to remember to take a camera and all the accessories with you, you might never take that picture. But if you have your Centrino laptop equipped with a camera, you don't need to lug a standalone camera at all.

Wireless networking takes the digital photography revolution to the next step.

With Wi-Fi, you can transmit your pictures for viewing anywhere from anywhere. This has implications in many different areas, including

- Data gathering for professionals: In a variety of professional fields, such as medicine, the ability to easily transmit photographs can be used in examinations, consultations, and for general communication.

- Teaching: The ability to instantly compare visuals with someone who is not present will be very useful in many educational contexts. For example, you could show a biology sample from your course lab work to a teacher at a distant location and get immediate feedback.

- Amateur photography: The ability to send and receive images seamlessly from a wide variety of devices will change the way people use photography. People will reach for a laptop computer using Intel Centrino mobile technology that also takes pictures in much the way they reach for pen and notepad today.

To start with, it's easy to save your pictures to the hard drive of the Centrino laptop used to take the pictures.

Using the Network Smart Capture application, select an image to be saved from the library of images, and click the Save As button, shown at the tip of the mouse cursor in Figure 5.12.

Next, give the photo a filename and select a folder for it, as shown in Figure 5.12. Click Save to save the picture.

The process works in the same way in the Scanner and Camera Wizard. With a picture selected, in the Picture Name and Destination pane, shown in Figure 5.13, you provide a filename and location (using the browse button).

> **NOTE**
> Network Smart Capture also provides easy ways to email photographs as attachments and to save them using image servers available to other computers on a wireless network.

FIGURE 5.12

It's easy to save a picture to your file system using Network Smart Capture.

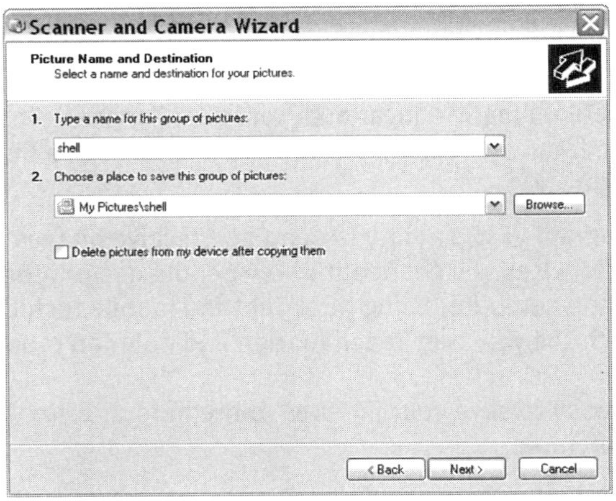

FIGURE 5.13

To save a picture, provide a name and location in the Scanner and Camera Wizard.

When the wizard completes, the picture will be saved with the name and location you designated (see Figure 5.14).

Whatever software you use to capture and save your photos, once they are part of your mobile computer's file system, you can use standard mechanisms for saving them to different machines on your wireless network. Windows Explorer is shown being used to copy a photo across my wireless network from one computer to another in Figure 5.15.

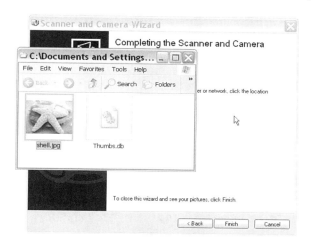

FIGURE 5.14

The file has been saved using the Scanner and Camera Wizard.

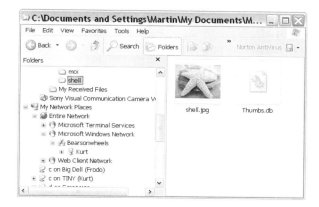

FIGURE 5.15

You can use Windows Explorer to copy a photograph from a mobile wireless computer to another computer.

You can also use email to send photographs as attachments and use standard mechanisms—such as FTP—to upload photographs taken with your Centrino laptop to the Internet.

Your Wi-Fi Photos on the Internet

You've taken your photos with your laptop using Intel Centrino mobile technology. You are connected to the Internet either at a hotspot (see Part III, "Mobile Computing on the Road") or through your own wireless network (as I explain in Part IV, "Your Own Wireless Network").

You'll find that many sites on the Internet allow you to do fun and useful things with your digital photos. These include

- Blogger, http://www.blogger.com, which lets you create a photo diary—or PhotoBlog—so you can share your pictures with the world

- Cardstore, http://www.cardstore.com, which lets you create and have printed greetings cards based on your photos (A fee is required.)

- Fotolog, http://www.fotolog.net, which lets you create photo blogs (This is probably the leading photographic blogging community.)

- ImageStation, http://www.imagestation.com, a service provided by Sony that lets you store, share, and print photos (Some services are free; some require payment.)

- Ofoto, http://www.ofoto.com, a service provided by Kodak that lets you store, share, and print photos (Some services are free; some require payment.)

Sending Photos with Email

If you want to email a photo to your friends or family, the easiest way to send the photo is as an email attachment. For example, if you use Outlook Express for your email program

1. Open Outlook Express.

2. Create a new email message.

3. With the message still open in Outlook Express, click the Attach button shown in Figure 5.16.

4. Use the Insert Attachment window to browse for the photo or photos you want to send, as shown in Figure 5.17.

5. After you have attached your photos, click the Send button to send your email. The photos you selected will be sent along with your message, as shown in Figure 5.18.

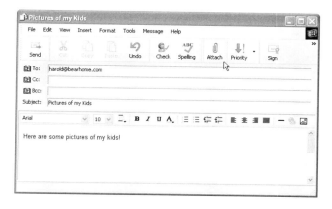

FIGURE 5.16

Click Attach to add photos to an email.

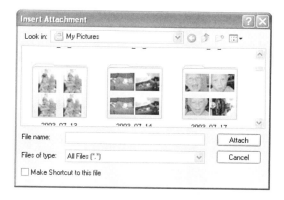

FIGURE 5.17

Using the Insert Attachment window, you can browse to select the photos you want to attach to an email.

Adding Your Pictures to Your Website

It's usually pretty easy to add the pictures you take with your mobile computer to your website, but the details of how to do so depend on how and where your website is hosted. Many website hosting companies provide special programs that make it easy to upload pictures (and other files) to your site. In any case, your web host has probably provided you with login information such as a name and a password.

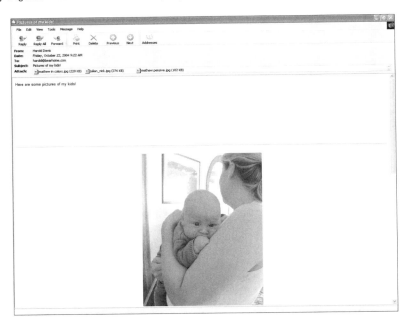

FIGURE 5.18

The photos attached to an email appear as part of the email massage when received in Outlook Express.

You can use your login information together with an FTP program to upload files directly to your web server.

You should bear in mind that to make your pictures available on the Web, you might want to place them within HTML pages and provide links to the pages. For the technically gifted among us, this is not a very difficult thing to do. However, the rest of us might want to enlist the aid of a visual web design program such as Microsoft's FrontPage or Macromedia's Dreamweaver.

Summary

Here are the key points to remember from this chapter:

- Digital cameras built in to mobile computers don't provide great picture resolution.

- Even though resolution isn't great, you can do many things with the pictures.

- Having a camera in your mobile computer means that you can carry one less gadget.

- It's easy and fun to take pictures with your laptop using Intel Centrino mobile technology if it has a built-in camera.

- You can use your wireless network to transfer pictures taken with your mobile computer.

- You can use a hotspot, or your wireless network Internet connection, to upload photos to many Web destinations.

Using Your Mobile Computer As a Telephone

I f you are traveling with your laptop that uses Intel Centrino mobile technology, and you are connected to the Internet at a hotspot, you can fairly easily use it as a telephone. As I'll explain later in this chapter, to do this, you'll need to be equipped with a microphone (unless there is one built in to your laptop) and a headset (unless you don't care about everyone around you hearing your conversation—in which case, you can simply use your computer's speakers).

You can also use your laptop as a telephone when it is connected to the Internet through your wireless network at home.

TIP

One problem with using your mobile computer as a phone is that the computer—and the related software—needs to be turned on for you to receive calls. However, an approach to getting around this, supported by some providers, is to have calls to your laptop forwarded to a cell phone or landline when your laptop is switched off.

A mobile computer is probably not as convenient for making or receiving phone calls as a normal cell phone. So what's the big deal?

Besides the fact that you don't have to lug multiple pieces of hardware around, by placing calls using your computer as your telephone, you can save a great deal of money on phone calls—and have fun while you are at it.

Understanding Internet Telephony

Internet telephony refers to routing voice telephone calls over an Internet Protocol, or IP, network. This section explains a little about how Internet telephony works. If all you want to do is learn how to use Internet telephony, you can skip ahead in this chapter to the sections that tell you where to sign up for the service, how much it will cost, and what additional hardware you may need.

Although a private network is involved in some cases, for the most part, phone calls are routed over the Internet. This is achieved using a combination of hardware and software.

Another term for Internet telephony, which I will be using in this chapter, is *VoIP*, short for Voice over IP.

Although the quality of VoIP may not be quite up to that of plain, old-fashioned telephonics, the quality is quickly getter better, and this has become less of a concern. In fact, most voice telephone calls you place today will in part be routed using VoIP, even if they start out (and end up) as regular voice telephone calls.

This brings up the point that VoIP comes into play a lot of ways in today's telephonics. For example, even if a phone call starts out being placed with a conventional "last mile" provider over old-fashioned phone line and received in an analogous fashion using a regular telephone, it is likely that the "long haul" portion of the phone call was routed using VoIP—simply because it is a less expensive way to manage telephonics than an old-fashioned dedicated network.

A more end-to-end VoIP solution is just starting to be offered by upstart telecommunication companies such as Vonage, `http://www.vonage.com`. Provided that you have a broadband Internet connection, Vonage proposes to replace your existing conventional phone system with VoIP-based telephones, which are both cheaper, easier to configure, and with more features—such as free multiline voice mail, call blocking, call forwarding, and more: all configurable telephonically or on the Web.

Pulver Innovations (`http://www.pulverinnovations.com`) makes a standalone Wi-Fi telephone that uses VoIP when it is in range of an open wireless broadcast signal (or one to which it can log on). There have been reports that Vonage is testing this phone for use with its VoIP service.

Skype offers VoIP service free anywhere in the world between computers that run Skype software. An extra-charge service called SkypeOut enables a PC to make calls to conventional phone numbers and another service called SkypeIn enables a caller on a conventional phone to call your PC. All these capabilities and more are described on the Skype website at `http://www.skype.com`.

Some companies are also introducing dual-mode cell and VoIP Wi-Fi phones. These use a cell network when Wi-Fi is unavailable and take advantage of the lower cost of VoIP over Wi-Fi when they can.

You can pretty easily turn your laptop that uses Intel Centrino mobile technology into a VoIP, Wi-Fi monster. Once you have done this, you can use your mobile computer to call other VoIP computers, or to call regular, old telephones as shown in Figure 6.1

Although carrying around a mobile computer that is also a telephone is bulkier than just carrying around a cell phone—is that a cell phone in your pocket, or are you just glad to see me—there are some real benefits to adding VoIP to your laptop. These include

- Only having to carry a single device (rather than two)

- Being able to take advantage of the lower cost of end-to-end Internet telephony

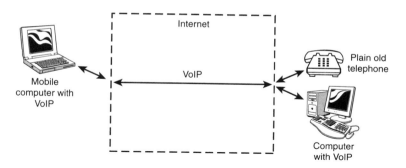

Figure 6.1

Calls are placed over the Internet from a computer using VoIP to another computer or a regular phone.

VoIP Vendors

Many vendors can supply the software and services necessary to add VoIP to your mobile computer.

Here are some of the most well-known VoIP vendors, along with a general idea of their charges:

- DeltaThree, http://www.deltathree.com, provides iConnectHere software and VoIP services with no-plan rates at about $.03 per minute worldwide (rates are lower with minimum purchase plans), and calls to other iConnectHere users are free altogether.

- Net2Phone, http://www.net2phone.com, provides VoIP software and services that is part of a package it calls CommCenter (see later in this chapter for more details). The portion of CommCenter most related to VoIP is called PC2Phone. Roughly speaking, calls cost $.02 per minute within the United States and $.03 worldwide.

> **NOTE**
>
> Essentially, the issues involved in adding VoIP to a mobile computer are the same as those involved in adding the capability to a desktop computer. In both cases, the computer needs to be connected via broadband to the Internet, whether via Wi-Fi to a hotspot, Wi-Fi through a private network, or via standard Ethernet cabling.

- Skype, http://www.skype.com, (mentioned earlier) based in Luxembourg, offers worldwide calling at about $.02–.03 per minute and free calling between computers using the Skype software.

- Vonage, http://www.vonage.com, also mentioned earlier as a leading provider of complete VoIP telephone systems, provides a software and service package called SoftPhone, which adds VoIP capability to your mobile computer or computers. The cost is an additional $10 per month for current Vonage subscribers and is not available to non-subscribers. (A typical Vonage plan costs $30 per month and features free calling within the United States.) One benefit of SoftPhone is that it works seamlessly with your other Vonage telephones.

- BroadVoice, `http://www.broadvoice.com`, has rate plans from $10 to $30 as well as a program to use the VoIP phone adapters from a variety of manufacturers. So, if you're not happy with one VoIP vendor, you can avoid equipment purchase costs when switching and use the same equipment with BroadVoice. BroadVoice soft phone support is available at $4 per month over the basic service. Telephone capability for your PC is provided by Microsoft Windows Messenger, a free download from Microsoft.

The Hardware You'll Need

As I've mentioned before, it's easiest to use your mobile computer as a telephone if you equip it with a microphone and headset. You don't have to do this: If your mobile computer comes with an internal microphone and a speaker, you can just use these. The downside is that everyone around you will hear both sides of your conversation. And the audio quality of your phone conversations might not be as good as with a dedicated headset and microphone. For one thing, background noises won't be screened out.

Most mobile computers are equipped with two connection jacks—one for a microphone and one for a headphones. If you look at your laptop, you'll likely find the microphone jack marked with an appropriate icon and the headset jack also marked as in Figure 6.2. (The microphone jack is usually color-coded red, and the headset jack is color-coded green.)

Your best bet is to probably buy an integrated headset microphone unit. These headset and microphone combos provide two plugs—one goes in the headset jack and the other into the microphone jack. They are primarily intended for telephony use with your mobile computer, although they work well for other applications, such as gaming and voice recognition. Headsets with microphones that fit in the jacks provided by the computer are called *analog* headsets and should cost between $25 and $50 for a good set.

You can also buy so-called *digital* headsets, which connect to your mobile computer's USB port and provide their own digital signal processing. Because these headsets do their own digital signal processing, they do not use the sound capabilities of your computer—a good thing if there is no sound hardware that comes with your computer, or if is just plain lousy.

Digital headsets cost a bit more than analog headsets, from about $50 to $150 depending on the make and model, but they probably deliver better quality.

A good place to buy either an analog or a digital headset, if you don't already have one, is `http://www.headsets.com`. Figure 6.3 shows the Headsets.com overview page for computer headsets.

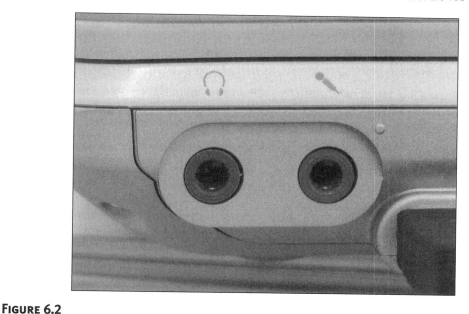

FIGURE 6.2

Your mobile computer provides jacks for a microphone and a headset.

FIGURE 6.3

You can buy digital or analog headsets for your computer at Headsets.com.

Bandwidth Considerations

Good bandwidth makes the difference between VoIP that sounds terrible and jerky (and is not worth it) and VoIP that sounds every bit as good as your regular telephone.

When you are looking at bandwidth for VoIP on your mobile computer, you need to think about two network segments (see Part IV, "Your Own Wireless Network," for more information about how these different parts of the network interoperate):

- The pipeline from the Internet to your access point or router

- The connection between your computer and the access point or router (At a public hotspot, this connection is provided by the hotspot's access point.)

There's not a whole lot you can do about the first of these, the Internet connection, other than try it and see. But as a ballpark, any decent broadband Internet connection should do.

The connection between your mobile computer and the gateway to the Internet (the access point or router) is a closer thing. Probably, 802.11b Wi-Fi is not good enough—but 802.11g is. In the wired world, 100BASE-T Ethernet will almost certainly do. (But 10BASE-T Ethernet is too slow.)

A useful way to test your connections for their capability of handling VoIP is the site provided by Brix Networks, TestYourVoIP, `http://www.testyourvoip.com`, shown in Figure 6.4.

TestYourVoIP allows you to run tests of connectivity speed for VoIP to a variety of locations. For example, as you can see in Figure 6.5, my VoIP connectivity from Berkeley, California to Helsinki, Finland is "better than a decent phone call," but not quite "as good as calling next door." In other words, it will probably do.

Installing VoIP Software and Making Your First Call

Generally speaking, it's easy to download and install the software that effectively turns your Centrino laptop into a telephone.

The steps are similar to those you follow anytime you download and install software from the Internet:

1. First, make sure that you trust the vendor of the software and that your antivirus software is up to date (see Chapter 17, "Protecting Your Mobile Wi-Fi Computer," for more information about antivirus software).

2. Next, download the VoIP software using the download link (or button) provided by the vendor.

3. Run the installation program provided by the software.

4. Finally, follow the registration process to open an account with the VoIP provider.

I'll show how this process works with two of the most popular VoIP providers, Net2Phone and Skype, and how to make your first call with each.

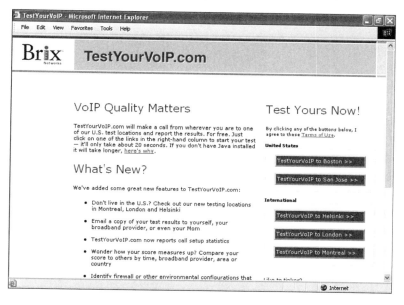

FIGURE 6.4

You can test your connection to the Internet for VoIP readiness using TestYourVoIP.com.

Net2Phone

First, open the Net2Phone site, and find the PC to Phone Telephone Service link, and then click it.

The Download link is on the extreme left; click it. You will be taken to the Download page for Net2Phone CommCenter. Click the Download Now button shown in Figure 6.6, and follow the more-or-less standard installation process. Once the software has been installed, you will be asked to initiate the registration process, which is different for new customers as opposed to returned customers (see Figure 6.7).

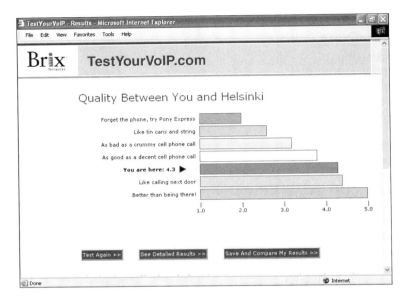

FIGURE 6.5

My VoIP connectivity between California and Finland will probably do for all but the most picky.

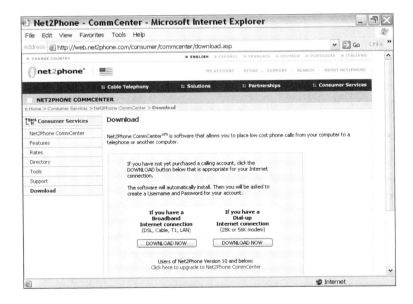

FIGURE 6.6

Click Download Now to start the Net2Phone installation process.

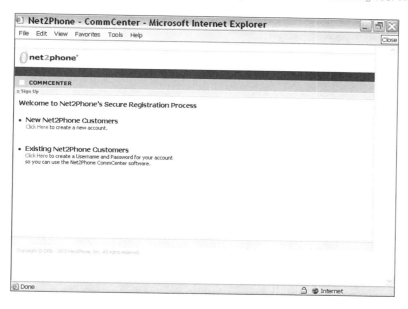

FIGURE 6.7

If you are a first-time user, you need to set up an account.

The information you provide to create an account is basically standard contact information, as you can see in Figure 6.8, and does not include credit card or other payment information.

With your registration information complete, you can now log in to the Net2Phone CommCenter. The first time you log in, the software will check the sound level of your speaker (or headset) and microphone.

When sound check is complete, you can place a phone call. As shown in Figure 6.9, this is simplicity itself.

Use the virtual keypad to enter the number you want to call, and click dial. You'll hear a dial tone, and then ringing, and then someone picking up the phone at the other end (that is, if someone is there).

> **TIP**
> You might want to be careful to uncheck the Yes, I Want to Receive Special Offers box shown at the bottom of Figure 6.8. (Who needs extra emails?)

A couple of things worth bearing in mind: Your user experience will be greatly improved if you use a headset rather than the speakers on your mobile computer. I mentioned a supplier of headsets earlier in this chapter; you can also buy them from Net2Phone.

Also, if you don't have any money "on file" with Net2Phone, your call will not go through until you have added some funds to your account using a credit card. So you might want to click the Add Funds button before you start placing calls.

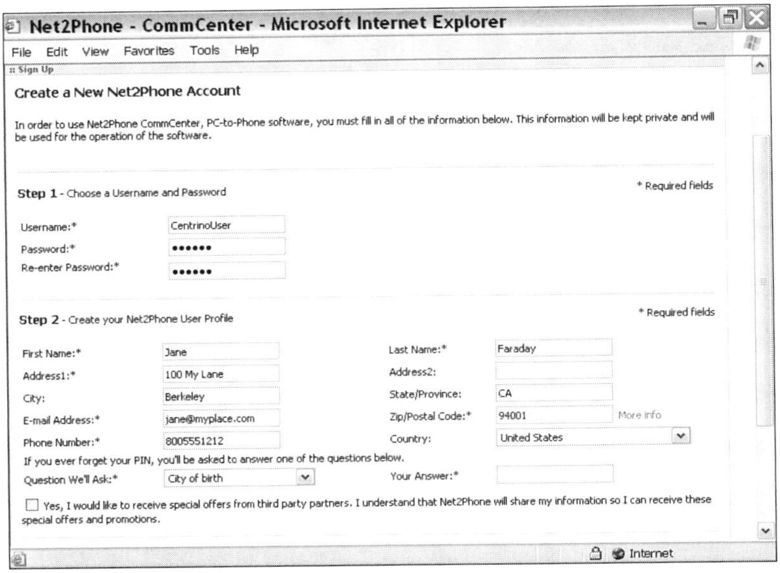

FIGURE 6.8

You must provide contact information to open a Net2Phone account.

FIGURE 6.9

To make a call, use the phone-like "keypad" to enter the number, and click Dial.

Net2Phone VoIP calling is free when the call is placed computer-to-computer to another Net2Phone user. The process works similar to instant messaging, and you initiate the process by clicking the Message button shown in Figure 6.9. To place this kind of "message," your counterparty needs to be up and running with Net2Phone, and you need to add him or her as a contact. (Click the Add button to add a Net2Phone user to your active contacts list.)

Finally, you might find the mobile computer to fax service provided by Net2Phone handy. In effect, this service means that there's really no need for you to maintain a separate fax machine because faxed documents can be retrieved electronically (at a very low per document cost).

Skype

Skype works in much the same way as Net2Phone, with paid (but cheap) international calling, and free contact with other Skype users. However, as opposed to Net2Phone, which is more or less a telephone service, Skype is promoting itself as a way of life along with file sharing and more conventional instant messaging.

Figure 6.10 shows the opening page of the Skype website, `http://www.skype.com`.

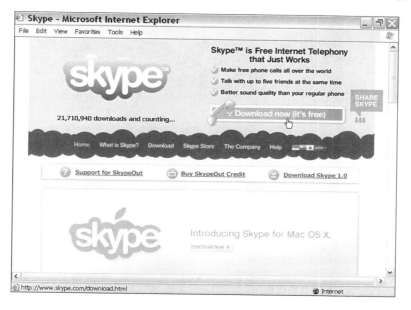

FIGURE 6.10

Click the Download button to get started installing the Skype software.

Click the Download button shown toward the upper right of Figure 6.10 to start the process of downloading and installing the software.

When the software has been downloaded, run the Skype installation utility.

After the software has been installed, you will be asked to create a Skype account, as shown in Figure 6.11.

FIGURE 6.11

When installation is complete, you will be asked to create a Skype account.

NOTE

You'll likely want to make sure that the Please Contact Me box shown in Figure 6.11 is unchecked. You should probably also uncheck the Start Skype When the Computer Starts box. Automatically starting VoIP software such as Skype poses a potential security risk—particularly if you forget that it is running.

After you have provided a Skype name and password, click Next. You will then be asked to complete an optional profile screen. If you plan to be an extensive Skype user, it is probably a good idea to supply at least some of this information.

Once you have updated your profile, the Skype window will open, shown in Figure 6.12.

To dial a call, click the Dial tab, shown in Figure 6.13.

You can enter a number to call using the virtual keypad shown in Figure 6.13, or by using your mobile computer's keyboard to enter the number directly in the drop-down list shown toward the bottom of Figure 6.13.

FIGURE 6.12

When you've updated your profile, the Skype window will open.

With the number entered, to place the call, click the telephone receiver icon toward the bottom left of the Dial tab. Your phone call will be placed.

Although Skype calls to other Skype computer users are free, as with the Net2Phone service, you cannot place a phone call to a conventional telephone number without first loading your Skype account with some money (to pay for the call). In Skype lingo, this is called "activating SkypeOut" by buying SkypeOut credit, as shown in Figure 6.14.

> **NOTE** Note that Skype expects you to enter a plus sign followed by a country code, even for local calls.

It's a great virtue of Skype that you can place a "call" to other Skype users free, but of course you must know their Skype contact name to be able to do this. If you don't already have some Skype friends, Figure 6.15 shows the search screen you can use to look for users who might want to talk with you.

FIGURE 6.13

The Skype Dial tab is used to place calls.

FIGURE 6.14

To place a call to a conventional phone number, you need to active SkypeOut, meaning fund your account.

FIGURE 6.15

You can search for other Skype users to talk with for free.

Summary

Here are the key points to remember from this chapter:

- Internet telephony—or VoIP—has become good enough so that you can viably use it for you phone calls.

- You'll need a microphone to place calls with your mobile computer.

- Although you can use the speakers built in to your mobile computer, your VoIP experience will definitely be happier if you purchase a headset.

- There are a number of competing VoIP software and service providers.

- It's free to download VoIP software.

- Telephone calls are inexpensive using VoIP.

- It's free to use VoIP to contact other users of the same VoIP software, and some VoIP vendors, such as Skype, are making an effort to foster a community of users.

> **TIP**
>
> If you are serious about finding a stranger to speak with using Skype, you probably should check the Search for people who are in "'Skype Me' mode," shown in Figure 6.15. Skype users who have this option turned on are affirmatively inviting calls.

Let Your Laptop Entertain You: Streaming Media, Gaming, and More

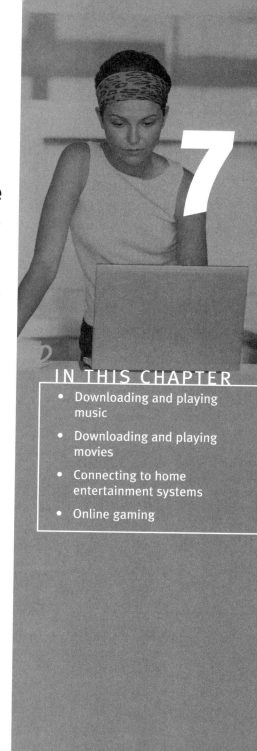

I f you've ever been bored—or ever find life too quiet— this chapter is for you. There's literally no end to the way your mobile computer can entertain you.

It's no secret today that you can use musical appliances such as the iPod to download and play your favorite music wherever you are. You can use your wireless laptop in the same way, and this chapter shows you how.

You can also use your mobile computer to play movies wherever you are. It's a great way to pass the time while waiting in an airport or on a crowded plane.

If the quality of picture or sound that you get from your mobile computer isn't good enough for you—and this will vary, of course, depending on your laptop and your personal tastes—you can play music and movies from your laptop to home entertainment appliances. In this chapter, I'll show you how.

A whole lot of online games exist that you can play with your wireless computer—for example, while connected to a public hotspot. I can't, of course, show you the whole universe of online gaming, but in this chapter, I'll point you to some of the best sites.

Downloading and Playing Music

It's easy to download and play your favorite music on your wireless computer, either from a public hotspot or from your home wireless network. Of course, sound quality will vary depending on your laptop. If the speakers that come with your laptop don't sound good enough for you—or you want to save your roommates from your loud and raucous musical tastes— you can always get headsets for your mobile computer.

As I'll show you later in this chapter, there are several ways you can stream music from your laptop to a home audio system via wireless. But for now, let's focus on finding and playing the music on a single mobile computer.

iTunes

Apple's iTunes, `http://www.itunes.com`, is probably today's largest single legal source of music for downloading, with more than 1,000,000 songs. Limited free previews, and free songs, are available, but for the most part each song costs $0.99. After you have purchased a song from iTunes, you are licensed to burn it on to a CD, to play it on an unlimited number of iPods, and to play it on up to five computers (both Apple and Windows).

> **TIP**
>
> Unlike some other music stores, the music available on iTunes can only be played using Apple's software.

To get started with iTunes, you first have to download the iTunes player, shown in Figure 7.1.

FIGURE 7.1

iTunes is used to download and play songs on your mobile computer.

With iTunes downloaded, click the Music Store link (shown on the left side of Figure 7.1) to start browsing the catalog of songs available for sample and purchase.

Music is categorized, so, for example, you can choose to browse Alternative or Country tunes (depending, of course, on your taste).

To purchase tunes, you'll need to establish an account. You can initiate the process of opening an account by clicking the Sign In button shown in the upper right of Figure 7.1. To establish an account, you will need to provide credit card, email, and physical address information.

With an account established, you can purchase a song for download by clicking the Buy Song button shown in Figure 7.2.

FIGURE 7.2

Click Buy Song to purchase a song.

After you've purchased and downloaded the song, you can play it using the controls at the upper left of the iTunes player.

Whether or not you are logged on to an Apple iTunes account, you can play samples of songs by selecting the song and clicking the play button located in the upper left of the iTunes player. For example, Figure 7.3 shows a sample from the song "Wayward Angel" being played.

MSN Music

Many other sites besides Apple's iTunes offer legitimate pay-for-download music including MusicMatch and Real Network's Rhapsody. One up-and-comer in this arena worth having a look at is MSN Music, located at `http://music.msn.com`, and shown in Figure 7.4.

FIGURE 7.3

You can play free samples by selecting a song and using the iTunes controls to play it.

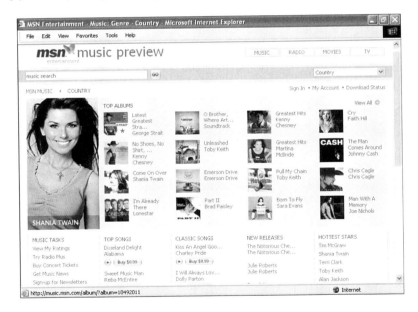

FIGURE 7.4

MSN Music is a good alternative to iTunes.

MSN Music is the new kid on the block and doesn't quite have the cool quotient of iTunes. Pricing is more or less the same as iTunes's—namely $0.99 per song.

A Microsoft Passport is used for logging on to MSN Music and paying for your purchases.

But MSN Music does have one important virtue that iTunes doesn't have—because the music is encoded in the WMA digital computer, it almost certainly can be played using the software already loaded on your mobile computer, so there's no need (as with iTunes) for an additional software download.

Using MSN Music, you can listen to streaming radio (which is something you don't need MSN Music for, as I'll show you in a little while). But the nifty feature is that if you use MSN Music to stream radio and hear something you like, you can buy and download it on the spot.

KaZaA

You probably also know that you can download songs from peer-to-peer file sharing networks such as KaZaA. To get started with KaZaA, you first need to download KaZaA from `http://www.kazaa.com`. KaZaA comes in two versions—one that is free and has ads and a paid version that costs $29.95.

> **TIP**
> As I explain in a moment, the free version of KaZaA installs spyware on your computer, so you should think twice about downloading it!

Music downloaded using KaZaA is free to download and can be played using the software already on your mobile computer. While you can download anything off KaZaA and other peer-to-peer networks for free, this doesn't mean that it is ethical, legal, or safe to do so—because the content on KaZaA may itself be copyrighted (see further discussion a little later).

As you might know, KaZaA is a peer-to-peer network, meaning that you are downloading music from other people. It is also very popular, with millions of downloads on any given week.

The fact the KaZaA and other similar service are peer-to-peer is both good and bad news. The good news, of course, is that downloads are free. The bad news is that there is no quality control on downloads, and they might even contain malicious software such as viruses and spyware.

The courts have ruled that it is perfectly legal to install KaZaA, and other peer-to-peer services, on your mobile computer. But it does not mean that everyone who makes music available for download using KaZaA has the right to do so. These downloads might not be authorized by the artists who created the songs, who also might never see any royalties from them. This means that by playing music you've downloaded off KaZaA you might be cheating the content creators of that music. Apart from ethical issues, the Recording Industry Association of America (RIAA) has launched enforcement lawsuits against individuals who download and play copyrighted music without permission. So there are risks involved in using KaZaA which may outweigh the "getting something for nothing" it provides. Since there are plenty of sites that let you download legitimate music for small fees (and you know you are not getting something extra like a virus or a lawsuit along with your download), why not stick to them and avoid KaZaA (and help an impoverished musician make a buck while you are at it)?

It's fascinating to explore the world of peer-to-peer networking. Whether you choose to use it to download music for your enjoyment, and under what conditions, is—like most things in life—up to you.

Streaming Radio

As I mentioned earlier, you can stream radio broadcasts to your mobile computer. This is very, very cool because using your connected wireless computer, you can listen to a streaming broadcast anywhere in the world.

It's pretty easy to search the Web for radio stations that provide their broadcasts as streams. You can also use a site that amalgamates a number of radio streams to make them easier to find. One example is MSN Music, which I mentioned earlier. Another is PublicRadioFan, `http://www.publicradiofan.com`, which provides an easy way to find public radio station streams.

Figure 7.5 shows the PublicRadioFan site with a RealPlayer stream playing jazz from WGBO in Newark, New Jersey. (I know you can't tell from the picture what's playing, but I thought I'd tell you so that you have some idea how good it sounds right now!)

The main PublicRadioFan window, shown in Figure 7.5, displays some of the most popular radio streams that are currently broadcasting. You can click the link for a program that is currently playing to stream it to your mobile computer.

TIP

You can see many, many more public radio streams by viewing the list of stations in the drop-down list provided by the Stream Launcher, shown at the right of Figure 7.5.

The icon to the left of the station indicates the type, or formatting, of the music stream. Most streams are formatted for use with RealPlayer, as MP3 files, or for Windows Media Player. Don't worry: You probably already have a player for each of these kinds of streams on your mobile computer; if not, you will be prompted to download the right player when you try to stream the sound file. (RealPlayer can be downloaded free from `http://www.real.com`.)

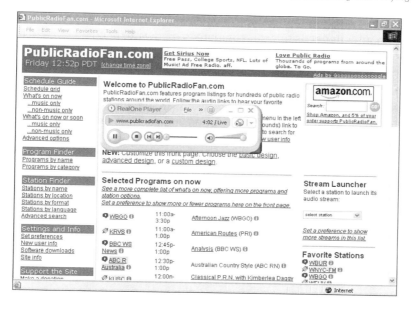

FIGURE 7.5

PublicRadioFan provides a mechanism for accessing the broadcast streams of many public radio stations.

Downloading and Playing Movies

You can download movies from the Internet, and enjoy them on your wireless mobile computer, in just about the same way you can download music. From a technical viewpoint, digital files are just digital files, and there is not much difference between movies and sound.

As with sound files, you need a player to view movie files. Both RealPlayer and Windows Media Player provide this facility—so it's likely you already have the right player on your laptop with Intel Centrino mobile technology. If you need to install this software, it is free—and you will be prompted through the installation steps if you need a media player to support a particular movie file (and don't already have the player you need).

Once again, as with sounds, how much you'll enjoy the movie experience on your mobile computer depends on your computer's hardware. I do recommend headsets, particularly if you are trying to watch a movie in a crowded environment. Keeping your movie's sound effects private will keep down ambient noise in general (so others will appreciate it) and will help you enjoy your movie experience.

In a moment, I'll show you a couple of websites that are in the business of providing movies for you to download (one on a pay as you go basis and the other using a subscription model). You should also know that movies are available for free download from peer-to-peer service such as KaZaA, discussed earlier in this chapter. (Once again, just because this content can be downloaded, it doesn't mean it is ethical or legal to use it without the permission of the copyright holders.) In other words, it's not just music that can be shared.

If you search the Web for sites that let you download movies, as you'd probably suspect, you'll find quite a few. One of the most popular is Movielink, `http://www.movielink.com`, a portion of whose catalog of movies is shown in Figure 7.6.

FIGURE 7.6

A portion of the extensive Movielink catalog is shown.

It costs from $1.99 to $4.99 to rent a movie from Movielink for a 24-hour period. The rental period starts when you first click Play, so you can download movies in advance, and your rental period won't start until you are ready to enjoy the movie.

MovieFlix, `http://www.movieflix.com`, is another popular movie download site. You can see a portion of the MovieFlix catalog in Figure 7.7.

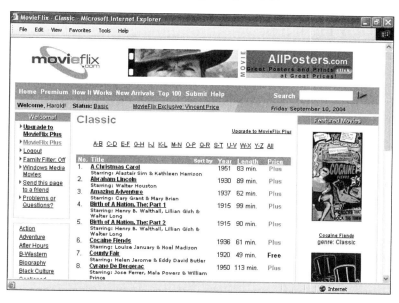

FIGURE 7.7

MovieFlix, a popular site for online movie downloads, has quite a bit of content available for free.

Unlike Movielink, MovieFlix works on a subscription model. For $6.95 per month you can download everything in the MovieFlix catalog. You'll also find quite a number of movies that are free for you to download at MovieFlix.

Another site that rents movies for download using a variety of pricing models is CinemaNow, `http://www.cinemanow.com`. CinemaNow makes a special point of stocking current movies, so if that's your interest, you might want to check it out.

Connecting to Home Entertainment Systems

You've used your wireless laptop to download that great song or movie, you listened or watched, and now you want to blast it over the house stereo or watch it on the big screen.

There are many ways to accomplish this.

It's possible to simply walk your mobile computer over to your stereo or television and make a connection using good, old-fashioned cabling.

A number of possible output connections on your mobile computer can be used to achieve this, including your USB connector. For a summary of the different possible ways to connect a mobile computer to a stereo, and the pluses and minuses of each, see `http://www.ramelectronics.net/html/howto-pc-audio.html`.

Connecting a mobile computer to a television or other home entertainment system via cabling is a little more complex than connecting to a sound system because video as well as audio signals are involved. This means that you probably will have to use the video output provided by your laptop, in addition to the cabling you'll need to connect the sound. For all the ins-and-outs involved in connecting a mobile computer to a television or home entertainment system, see `http://www.ramelectronics.net/html/howto-pc-tv.html` and `http://www.cinemanow.com/PC-to-TV.aspx`.

But who needs all those wires? The theme of *Anywhere Computing with Laptops: Making Mobile Easier* is that anything you can do with wires you can do better without them.

If you want to connect to the video portion of a home entertainment system as well as the audio portion, you should investigate a category of device called the *media receiver*, also sometimes called a media adapter.

Media receivers let you share music and video wirelessly with stereos and home entertainment systems. The media receiver has outputs for audio and video and sits near your stereo or home entertainment system. It connects wirelessly to the signal broadcasting from your access point (see Part IV, "Your Own Wireless Network," for more information about how this works).

The media receiver scans your network for computers running its software, such as your mobile computer. Once found, you can pick and choose what the media receiver is to play on the stereo or television. As an added bonus, these devices come with remote controls, so you can select your content without being near either your computer or the entertainment center.

One of the most popular media receivers is the EZ-Stream Universal Wireless Multimedia Receiver from SMC Networks, which uses both the 802.11a and 802.11g flavors of Wi-Fi, and costs between $150 and $200.

Online Gaming

Games! Games! Games! Everyone loves games, and there's nothing better than a game you can connect to using wireless from your favorite hotspot or from your private wireless network.

Here are the some of the best sites to use to find online games:

- Games.com, `http://www.games.com`, provides board and word games, arcade games, and card and dice games. Some are solo, and some involve online interaction. Playing most of these games is free.

- Jolt, `http://www.jolt.co.uk`, is Europe's most popular online gaming network, featuring games such as Evil Genius—in which (you guessed it) the goal is to conquer the world from your secret base using super secret agents.

- MMORPG, `http://www.mpog.com`, features, per its acronym, "Massive Multiplayer Online Role-Playing Games" such as *The Saga of Ryzom* (see Figure 7.8).

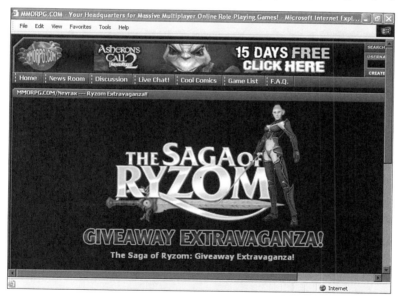

FIGURE 7.8

MMORPG, or Massive Multiplayer Online Role-Playing Games, features role-playing extravaganzas such as The Saga of Ryzom.

- Multiplayer Online Games Directory, `http://www.mpogd.com`, is—as the name implies—a directory of online games that are intended to be played by multiple people at the same time. Fun, fun, fun!

- The Online Gaming League, `http://www.worldogl.com`, provides a mechanism for putting together teams, or "tribes," of online game enthusiasts.

- Playsite, `http://www.playsite.com`, provides mostly interactive games. The site features both free and for-pay games and includes some gambling.

- Pogo, `http://www.pogo.com`, offers free and subscription content, primarily of a family-oriented nature.

- Station, `http://www.station.com`, offers online interactive versions of many of the Sony PlayStation games.

So, have fun with all these games and your mobile computer that uses Intel Centrino mobile technology. Whatever your gaming predilections, you're sure to have a great time.

Summary

Here are the key points to remember from this chapter:

- You can download music to your wireless computer from many online sources.

- It's easy to play the music you've downloaded.

- Many online resources exist with movies available for download.

- You can connect your mobile computer to a stereo or television using conventional cabling.

- Wireless connectivity is the really cool way to connect your mobile computer to a home entertainment system.

- If online gaming is your thing, you'll find games for every taste, ready for you to enjoy from a public hotspot or your private wireless network.

MOBILE COMPUTING ON THE ROAD

PART III

Entering a World Without Wires

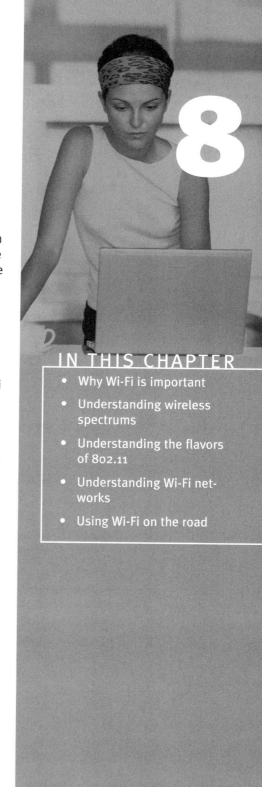

As you probably know by now, with your mobile computer, you can surf the Internet without wires from remote locations such as poolside at luxury hotels, in airports, in coffee shops, and in bookstores. You can also use wireless technology to set up networks in your home or office that allow you to share files, to share a connection with the Internet, and much more.

The underlying wireless technology that makes this all work is popularly called Wi-Fi, which is short for *wireless fidelity*. You don't have to understand a whole lot about Wi-Fi to be able to make good use of it. Therefore, this chapter does not go into a great deal of technical detail about the various Wi-Fi standards. You'll find some of the gory, geeky details in Appendix A, "Wireless Standards," if you really want to know.

Still, there are a number of good reasons for you to have some basic understanding of how wireless technologies work and the alphabet soup of standards they use. A little information will help you make knowledgeable decisions when you buy equipment. Being educated will also make it easier for you to get rolling with your mobile Centrino computer to start working wirelessly.

This chapter tells you everything you really *need* to know about the wireless standards related to Wi-Fi. It also provides an overview of the two main conceptual issues that are likely to confront the novice user of wireless technology: how wireless networks work and how to connect to wireless hotspots on the road.

Understanding Wi-Fi

The very short version is that Wi-Fi is a way for wireless devices to communicate.

Wi-Fi is the Wi-Fi Alliance's name for a wireless standard, or protocol, used for wireless communication. I'll tell you a bit more about this wireless standard and its variations, known collectively as IEEE 802.11, later in this chapter. (IEEE stands for the Institute of Electrical and Electronics Engineers, which defines the standard.)

THE WI-FI ALLIANCE

The Wi-Fi Alliance is a not-for-profit organization that certifies the interoperability of wireless devices built around the 802.11 standard. The goals of the Wi-Fi Alliance are to promote interoperability of devices based on 802.11 and, presumably, to promote and enhance the standard.

For better or worse, this is no neutral organization. The members of the Wi-Fi Alliance are manufacturers that build 802.11 devices. As of this writing, more than 200 companies belong to the Wi-Fi Alliance and more than 1500 products have been certified as Wi-Fi interoperable.

The Wi-Fi Alliance makes the promise that if you buy an 802.11 device with the Wi-Fi seal of certification, the device will work seamlessly with any other Wi-Fi certified device—at least those running the same flavor of 802.11.

You can find more information about the Wi-Fi Alliance at the Alliance's website, `http://www.wi-fi.org`.

Standards and protocols are mostly of interest to engineers (however, see the sidebar "What Is a Standard?" for more information if you are curious).

But Wi-Fi has garnered a huge amount of attention from people who would normally be unconcerned about engineering details: in other words, normal human beings such as you and me. Students, professionals, homemakers, Comparative Lit majors, and office workers are all talking about Wi-Fi.

The really big question is: Why is Wi-Fi getting all this attention? I'll get to that soon. I'll also show you how Wi-Fi can change your life. (For real!) But first, I'd like to tell you a little bit more about what Wi-Fi is.

For now, you need to know that Wi-Fi devices are certified interoperable (within certain limits that I'll discuss shortly) and run on some flavor of 802.11, a medium-range wireless networking standard. It's important to know that 802.11 runs at speeds fast enough to handle network traffic (perhaps a little slower than a wired network). I'll be telling you about transmission speeds in more detail later in this chapter, but the crux of the matter is that if you are connected via Wi-Fi to the Internet (either through a private network or a hotspot), the Wi-Fi connection will not be the bottleneck, if there is one.

WHAT IS A STANDARD?

The words *standard* and *protocol* are essentially synonymous. (Protocol is a slightly more technical term.) When used in its engineering context, a standard means the technical form of something such as a message or a communication. In other words, a standard might specify how the communication is made.

If you know the standard, you know how to decode the message. In order to work with a standard (called *complying* with a standard), a device needs to know both how to encode into the standard and decode from the standard.

A standard for working with communications, such as the Wi-Fi standard, will generally involve specifications both at the hardware and the software level (in geek speak, these levels are called *layers*).

You can think of the standard as a kind of secret handshake that gets you into a club. If you (or your wireless device) know how the secret handshake works, you can find out what the other people in the club (the other wireless devices) are actually saying.

Wireless Spectrums

All you might ever want or need to know about wireless spectrums is that they are used to send and receive signals by your mobile computer, cell phone, and garage door opener. If anything more than this makes your eyes glaze over, please skip this section. But otherwise, if you are curious, please read on.

As you probably know, any signal sent without wires is called a *radio transmission*. A common example is that the radio in your car receives transmissions. Similarly, a standard cell phone works by receiving—and transmitting—radio signals.

Every device that broadcasts a radio transmission does so at a particular *frequency*, which is the oscillations, or movement from peak to trough, of the electromagnetic wave created by the transmission.

The entire set of radio frequencies is known as the radio *spectrum*. Contiguous portions of the radio spectrum are called *bands*, as in "the FM band."

Radio frequencies describe the oscillations of a radio wave. For example, if you are tuned to an FM radio station at 92.5, it means that the radio transmission is oscillating at 92.5 megahertz per second. 92.5 megahertz (pronounced "may-ga-hurts" and abbreviated MHz) means that the radio transmission wave oscillates, or moves from its valley to its peak, at a rate of 92,500,000 times per second. If you think of this as listening from a distance to a really rapidly vibrating tuning fork, you have the right picture.

The AM radio spectrum was developed before the FM spectrum, so it is *lower* down the spectrum, ranging from 535 kilohertz to 1.7 megahertz, or 535,000 to 1,700,000 oscillations per second. For example, 720 on the AM dial means that your radio receiver is tuned to a frequency of 720,000 oscillations per second.

There are frequencies *above*, or higher than, the FM frequency as well as below it. In fact, as I'll explain in a moment, Wi-Fi transmissions run at some of these higher frequencies.

> **NOTE**
> You don't really need to know much about frequencies to understand Wi-Fi. But if you are curious, for historical reasons, lower frequencies have been developed and used sooner than higher frequencies. You might be interested to know that visible light oscillates on the electromagnetic frequency at a higher frequency than any used for radio transmissions.

One thousand megahertz is equal to one gigahertz (pronounced "giga-hurts" and abbreviated GHz). So when you refer to the 2.4GHz frequency that Wi-Fi uses, you are actually talking about 2,400,000,000 (2.4 billion) oscillations per second.

Only so many frequencies in the radio spectrum can be used for transmissions. This has inevitably led to the potential for conflicts about usage, as well as attempts to dominate particular frequencies.

As a partial answer to frequency conflicts, the government has regulated the usage of most of these frequencies. In the United States, government regulation of radio frequencies is controlled by the Federal Communications Commission (FCC).

Some frequencies are reserved for particular usages, such as the military. Others, such as the AM and FM bands, are licensed. This means that only the licensees can use the frequency for the purpose it was licensed. In addition, some areas of the spectrum have been set aside for unlicensed, or "free," uses. These set-aside areas include the 2.4GHz and 5GHz spectrums, which is what Wi-Fi uses, as I'll explain later in this chapter in "The Free Spectrums."

> **TIP**
> If you find that your Wi-Fi device is getting interference from some other appliance such as a microwave or wireless telephone, one of the first things to try is moving either your Wi-Fi device, or the other device, to a new physical location.

The uses of some of the frequencies in the radio spectrum are shown in Figure 8.1.

Wi-Fi Standard Layers

The 802.11 (and Wi-Fi) standard includes what is called a *physical* layer. This physical layer uses something known as Direct Sequence Spread Spectrum technology (DSSS) to prevent collisions and avoid interference between devices operating on the same spectrum. You'll find much the same kind of technology in your wireless telephone handset. The idea here is that you don't want the signal coming out of your microwave unit to interfere with your email (or vice versa).

In addition to its physical layer, each 802.11 Wi-Fi device has an *media access control* (MAC) layer. The MAC layer specifies how a Wi-Fi device, such as a mobile computer, communicates with another Wi-Fi device, such as a wireless access point.

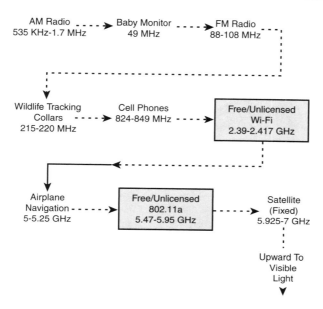

FIGURE 8.1

Selected uses for the radio spectrum (not drawn to scale; source U.S. Department of Commerce).

Together, the physical and MAC layers, along with extensions intended to implement extra features (such as security), make up the 802.11 Wi-Fi standard.

The Free Spectrums

Unlike many other wireless standards, 802.11 runs on "free" portions of the radio spectrum as I just explained. This means that (unlike cell telephone communications) no license is required to broadcast or communicate using 802.11 (or Wi-Fi).

The fact that the 2.4GHz and 5GHz frequencies have been set aside for unlicensed usages does have an extremely important implication: They are cheap to use. This gives these "free" spectrums an unfair competitive advantage compared to using a spectrum that someone has paid for. But there are some legal restrictions on what you can do within the free spectrums.

The free portions of the radio spectrum used by 802.11 (and Wi-Fi) are the 2.4GHz band, and, more recently, the 5GHz band. As you might know, many household appliances such as microwave ovens and (most significantly) wireless telephone handsets also use these free spectrums.

With a wireless telephone handset, a base station is connected to the telephone line, and the handset communicates with the base station over the "free" radio frequency, so that you can roam about your home or office while talking on the phone. Clearly, these wireless telephone handsets are not the same thing as cell phones, which do not connect to a telephone wire at all and use licensed portions of the spectrum.

> **NOTE**
> There are conflicts within the 5GHz band as well as the 2.4GHz band, but less so because the 5GHz band conflicts primarily concern competing usages such as radar and satellite radio, which are being ironed out by the FCC.

In other words, the 2.4GHz spectrum has become like a shanty town in which it is cheap to live. All kinds of transmission devices have crowded into the neighborhood, from microwaves to cordless telephones. These devices can interfere with your Wi-Fi transmissions and reception. In Part IV, "Your Own Wireless Network," I'll show you how to best avoid problems with competing 2.4GHz devices when setting up a Wi-Fi network.

Interference, of course, is a two-way street. You also don't want your Wi-Fi network wrecking havoc on your cordless phones and other devices that use the spectrum.

802.11 and Its Variations

Generally, the core 802.11 standard is intended to specify a way for computers to network using the 2.4GHz and 5GHz free spectrums I just explained. (When computers network, it is said that they are forming a *local area network*, or LAN. When computers network wirelessly, it is called a Wireless LAN, or WLAN.)

The 802.11b Standard

The 802.11b subset of the general 802.11 standard is yesterday's version of Wi-Fi, but it is the wireless networking technology still most generally in use and the least expensive technology you can buy today. (By the time I write the second edition of this book, 802.11b will probably be history, 802.11g will probably be the inexpensive technology, and newer standards will be cutting edge!)

The full 802.11b specification document is more than 500 pages long, but here are the key things to know about 802.11b:

- The 802.11b standard uses the 2.4GHz spectrum.

- The 802.11b standard uses a technology called Direct Sequence Spread Spectrum (DSSS) to minimize interference with other devices transmitting on the 2.4GHz spectrum.

- The 802.11b standard has a theoretical throughput speed of 11 megabits per second (Mbps).

The 11Mbps speed compares favorably with the 10Mbps throughput of a conventional 10BASE-T wired Ethernet network, although of course it is slower than the 100BASE-T network you are likely used to (which run at 100Mbps). It is certainly faster than even the fastest broadband Internet connections.

However, for a variety of reasons, Wi-Fi connections rarely achieve anything like its theoretical maximum. (Encryption slows 802.11b down, for one thing.) Weak connectivity also slows Wi-Fi down. Even so, Wi-Fi connections should be fine for everyday uses such as file sharing or sharing an Internet connection. Certainly, if you are planning to use more demanding applications, such as telephonics or media streaming, you should look to a faster standard, such as 802.11g.

I'll be telling you a little more about transmission speeds of 802.11b related to other wireless standards later in this chapter.

The 802.11a Standard

The 802.11a standard is a slightly more recent version of Wi-Fi that is somewhat faster than 802.11b. The 802.11a standard uses the 5GHz band for transmission, which minimizes the possibility of interference from the many 2.4GHz devices out there (think microwaves, garage door openers, and so on) and promises a theoretic throughput of 24MBps.

The 802.11a standard poses some compatibility issues with 802.11b. This means that unless you plan to only use your Centrino laptop on a private 802.11a network, if you do buy one that uses 802.11a, you should also make sure that it runs 802.11b and hopefully 802.11g as well.

The chief advantage of 802.11a is that it will run in to less interruption from other devices because it does not use the crowded 2.4GHz band. In addition, an advantage for 802.11a in an enterprise setting is that it has 8 channels (versus 3 channels in 802.11b or 802.11g). This is a potentially huge performance booster for the busy enterprise with a large number of users.

The 802.11g Standard

The gold standard of current Wi-Fi devices is 802.11g, which operates on the 2.4GHz spectrum and boasts throughput as fast as 54Mps.

The 802.11g standard is a little faster than 802.11a and much faster than 802.11b (see "Transmission Speeds" later in this chapter for further comparison). It is also inherently backward compatible with 802.11b—although 802.11b devices on an 802.11g network can slow down the 802.11g computers (with the amount of performance degradation depending on many factors).

802.11g is comparable to 802.11b, only faster, and compatible with 802.11b. Except for a very small incremental cost, there is no reason not to invest in 802.11g, the more up-to-date technology.

Centrino and 802.11

Almost all new Centrino laptops on the market today have the Intel PRO/Wireless 2200BG Network Connection (802.11b/g) Wi-Fi card as standard equipment. High-end Centrino laptops will have the all-singing, all-dancing Intel PRO/Wireless 2915ABG Network Connection (802.11a/b/g). So you really don't have to worry much about the performance of the wireless networking in your new Centrino laptop—it will be great with either Intel wireless interface. It probably won't be a problem, but just to be on the safe side, check to see that your new Centrino laptop has at least the 802.11b/g card.

Intel has other PRO/Wireless Network Connection products that are no longer in common use. There is the PRO/Wireless 2100 Network Connection that only supports 802.11b. Then there is the PRO/Wireless 2100A Network Connection supporting 802.11a and 802.11b. You should not see either of these cards on any current Centrino laptop for sale today.

There really is no good reason to buy 802.11b-only equipment of any type. The price of 802.11b/g equipment (internal wireless cards and access points) is about the same as 802.11b only. And who really wants to go at only 11Mbps when you can network instead at 54Mbps?

Related Wireless Standards

There's a constant process in which industry groups, sometimes competing, develop and propose 802.11 standards for approval by the IEEE. In time, the Wi-Fi Alliance usually adapts and certifies subsets of these standards in what amounts to a Darwinian process—the standards for which there are actual, real-world uses, and that form the basis for actual hardware, get certified.

A security standard for 802.11 of this sort is named 802.11i. The Wi-Fi Alliance has released a subset of the 802.11i standard that the Alliance has developed called *Wi-Fi Protected Access*.

Products that successfully complete the Wi-Fi Alliance testing required for meeting its version of the 802.11i standard are labeled as Wi-Fi Protected Access certified.

Wi-Fi Protected Access provides encryption and authentication services and is software implementable. For more about Wi-Fi and security, see Part V, "Securing Your Computer and Network."

Another security standard is 802.1x. This standard specifies a method for automatically revoking and providing new encryption keys.

Just as 802.11g is faster than 802.11b, a new standard, 802.11n, is expected to have speeds on the order of 100Mbps, almost twice as fast as 802.11g. The fly in the ointment is that the details of 802.11n are still in flux, with different industry groups proposing different implementation details.

WiMAX, based on the IEEE 802.16 standard, is primarily supported by Intel Corporation, and promises both the speed and the range to address the last mile problem of providing broadband networking over wide geographic areas. See "The Future Is Faster, Wireless Is Everywhere" in Chapter 1, "Understanding Intel Centrino Mobile Technology," for more information on the role WiMAX probably will play.

Two other wireless standards you might hear about are Bluetooth and 3G.

Bluetooth is a short-range connectivity solution designed for data exchange between devices such as printers, cell phones, and PDAs. Like 802.11b, it uses the 2.4GHz spectrum. Although Bluetooth is built in to a great many devices, it is a standard with some severe disadvantages, mainly that it is far slower than 802.11b (with nominal throughput of up to 721 kilobytes per second) and with a maximum range of about 30 feet (compared to Wi-Fi's unamplified range of several hundred feet). Bluetooth's main claim to fame is that it is inexpensive, which is why it has been added to so many devices. So, as you can see, Bluetooth is not really comparable to 802.11 wireless networking standards.

3G is a catchall term for a proprietary network based on cellular phone technology using spectrums leased by telecommunications carriers such as Sprint and Verizon. As opposed to Wi-Fi, it uses a leased spectrum that is not free for the telecommunications companies, and users must pay for it.

At this point, a few telecommunications carriers (notably Verizon and Sprint) are starting to offer 3G services for some of their coverage areas—at a cost starting at about $90 per month. These services offer speeds on the order of 100 kilobits per second (Kbps), or about 1,000 times slower than 802.11b-based Wi-Fi. 3G services are faster than 56Kbps dial-up and slower than Wi-Fi and are expensive, so they probably are not for the average person today.

Transmission Speeds

Figure 8.2 shows some comparative throughput speeds for most of the wireless standards I've discussed in this chapter.

To keep the comparison real world, I've included estimated actual throughput for 802.11b and 802.11g as well as their theoretical maximums.

For purposes of reference, and to help you see that Wi-Fi is unlikely to slow you down much, I've included the throughput you can expect from a 10BASE-T Ethernet network in the figure (the slowest wired network you'll encounter) and 100BASE-T Ethernet, the flavor of wires network most commonly in use.

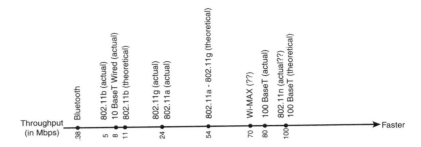

(Not drawn to scale)
All speeds appx and subject to real-world variables
?? = subject to confirmation in the real world

FIGURE 8.2

Comparative speeds of wireless standards (not drawn to scale).

You should take away these points from Figure 8.2:

1. 802.11g Wi-Fi is a little slower in the real world than a fast wired network, but more than adequate for office and home networks.

2. Even 802.11b is fast enough for most uses (and faster than your Internet connection).

3. Wireless networking is getting faster and better all the time.

Understanding Wi-Fi Networks

You're on the road, and you've found a location with a Wi-Fi broadcast device that your mobile computer can talk to. A Wi-Fi broadcast device is variously referred to as an *access point*, an *AP*, or a *hotspot*.

With your access point located, you're ready to sit right down, establish a wireless connection, and start reading your email and surfing the Web, right? Not so fast, partner!

It's really important to understand that being able to "talk" with a wireless access point just means that you can "talk" with a wireless access point. It doesn't mean that you can connect to the Internet unless the wireless access point is itself connected to the Internet.

So if Starbucks (or whoever) wants to provide you with the chance to surf on their turf while you sip that latte, Starbucks needs to provide an Internet connection. Generally, this connection is wired, and uses a cable or DSL (digital subscriber line) telephone for high speeds.

A high-speed wire brings the Internet to the location, and a Wi-Fi access point broadcasts the wireless Internet connectivity to wireless devices. (In technogeek speak, the wireless devices are generically referred to as *clients*.)

Between the Internet connection and the Wi-Fi access point, some hardware also needs to be designed to connect with the Internet and share the connectivity. This can be done a whole lot of different ways, depending on many factors. For example, is a wired network also involved? I'll be getting into these details in Part IV.

For now, you need to understand that connecting to the Internet via Wi-Fi involves four things:

1. Your Wi-Fi device (the client)

2. A Wi-Fi broadcast unit (the access point)

3. Network connectivity hardware (such as a router and modem)

4. The actual Internet connection (usually via cable or DSL)

A fairly typical simple Wi-Fi network setup of this sort, that lets Wi-Fi users connect to the Internet, is shown in Figure 8.3.

Hitting the Road with Wi-Fi

The rest of Part III, "Mobile Computing on the Road," includes a lot of detailed tips, tricks, techniques, and information about connecting using your laptop with Intel Centrino mobile technology and Wi-Fi when you are on the road. This section introduces the basics related to using Wi-Fi on the go.

For starters, as I explained in the preceding section, "Understanding Wi-Fi Networks," it's not enough to find a Wi-Fi wireless access point. Behind the scenes, the wireless access points need to be capable of providing access to the Internet, usually via a high-speed cable or DSL connection.

Figure 8.3 shows you a pretty typical example of how this might work behind the scenes.

Connecting

With your Wi-Fi device happily chugging and ready to go, a good strong signal from a Wi-Fi access point broadcasting its way to you, and a behind-the-scenes Internet connection that the access point is plugged in to, what's next?

Connecting to a public Wi-Fi access point is generally really easy. Usually all you have to do is open your Web browser. In some cases, any Web browser will do, but often you need to use Microsoft's Internet Explorer. Connecting is really quite simple, but I'll be showing you an example of how this works in Chapter 9, "Finding Hotspots."

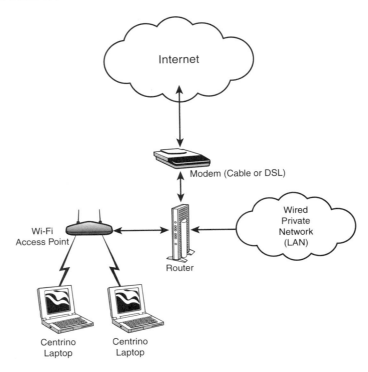

FIGURE 8.3

To connect to the Internet via Wi-Fi, the wireless Wi-Fi access point must be plugged in to equipment and an Internet connection, usually cable or DSL.

Paying for It

There are two sides to Wi-Fi. One is the grassroots peoples' movement in which propeller-headed persons (such as the author of this book) put up free Wi-Fi access points that can be freely used by anyone within range. Of course, in my case the range is about 50 feet from my house in a quiet residential neighborhood, so it might help a neighbor or two, but is unlikely to be of much use to anyone else. Seriously, if you want to offer free Wi-Fi using your Internet access, you should take care to set up a network that uses hotspot architecture and protects your private network resources.

The capitalistic, entrepreneurial side of Wi-Fi is what you might meet when you access the Internet via Wi-Fi in public venues such as Starbucks, Borders, hotels, and airports. In some cases, there are no free lunches in these venues! You get charged for wireless access.

On the other hand, some businesses have found that offering free Wi-Fi is a great way to entice customers to stick around for a while and spend more money. In addition, there is a hard-core contingent of idealistic engineering types who see offering free Wi-Fi access as a worthy endeavor. So perhaps sometimes there are free lunches! And when there are, they are really tasty!

Typically, you pay for your access based on how long you use it either with a payment plan, or as you go. If you plan to pay as you go, you'll need to have a credit card handy when you log on to the Wi-Fi network.

Finding Access Points

Where are the public Wi-Fi access points? Why, everywhere and nowhere, like a taxicab on a rainy day.

Actually, the picture is not nearly so bleak, and there are more and more wireless access points every day.

There are also a great number of online tools that help you find access points that meet your needs. Chapter 9 provides a great deal of information about tools you can use to find hotspots, and Appendix B, "Where the Hotspots Are," gives you some specific (and hard to find elsewhere) information about hotspot locations.

Summary

Here are the key points to remember from this chapter:

- All radio transmissions operate on a spectrum band.

- Wi-Fi uses the unlicensed 2.4GHz and 5GHz bands.

- 802.11 is the engineer's name for a wireless standard that uses a free portion of the broadcast spectrum.

- Wi-Fi is the name given to wireless devices that are certified to be compatible and use the 802.11 standard.

- 802.11g is rapidly becoming the predominate flavor of Wi-Fi today. 802.11g equipment costs about the same as 802.11b, but 802.11g is five times faster.

- Because 802.11g is backward compatible with 802.11b, and only a little more expensive, if you are buying a laptop that uses Intel Centrino mobile technology, you should be sure that it runs 802.11g.

- Wi-Fi provides data throughput that is fine for most uses.

- This book explains how to connect your computer using Intel Centrino mobile technology via Wi-Fi on the road, as well as how to set up a Wi-Fi wireless network at home (or in the office).

- Wi-Fi wireless Internet access requires more than just a Wi-Fi access point. To access the Internet, you'll also need an Internet connection and an intermediate layer of hardware equipment.

Finding Hotspots

You're armed and ready with your laptop that uses Intel Centrino mobile technology, and you want to get out on the town and surf the Internet. Or perhaps you are about to go on a trip—for business or pleasure—and want to make sure that you can be in touch with email using Wi-Fi every step of the way. Or maybe you simply want to walk your neighborhood and see if anyone has a Wi-Fi network you can use.

One thing that is great fun to do with Wi-Fi is to connect to the Internet from a remote location such as a hotel lobby or a coffee shop. If you are on a business trip, this kind of anywhere computing is not only fun and convenient, but it can help you be more productive by using your time more efficiently. The ability to connect via wireless on the road—for example, while waiting at an airport—lets you pick up email, and surf for information, when and where you need it.

The good news is that there is no great trick these days to connecting remotely via Wi-Fi. It's often as easy as turning on your mobile computer and opening a Web browser.

This chapter shows you how to hit the road with Wi-Fi, even if the road is only your nearby Borders bookstore or Starbucks latte emporium.

Where Are Hotspots Likely to Be?

The answer—at least this is quickly becoming the answer as Wi-Fi gets added to fast food joints, hotel chains, airports—is that Wi-Fi is everywhere. Being able to connect to the Net wherever one wants will soon seem as natural as being able to breathe. To my mind, the right to surf without wires will rank right up there with life, liberty, and the pursuit of happiness (not to mention the right to chug down the freeway in a giant-size gas guzzling vehicle).

You can find Wi-Fi in all kinds of unlikely places (see the sidebar "Mobile Computing Anywhere" for an example).

Nevertheless, it can take some effort to find Wi-Fi hotspots—that is, places that allow you to connect to the Internet with your laptop that uses Intel Centrino mobile technology—right where you want and need them.

In some places, it is pretty easy. For example, where I live, in Berkeley, California, there are literally hundreds of Wi-Fi hotspots. If I take out my laptop and walk in any direction, I can't go very far without hitting a place that provides Wi-Fi access.

It's a pretty safe bet that you'd find it hard to connect using Wi-Fi high up in the mountains of Wyoming. If the only occupants for hundreds of square miles are sheep, coyotes, and an occasional grizzly bear, no one will have thought to provide the infrastructure to let you connect via Wi-Fi. (Of course, as time goes by, the hibernating bears might feel that they've got to have those hotspots to keep from going crazy with boredom during the long winter nights.)

A little less fatuously, if I listed the absolute minimum requirements for Wi-Fi, I'd say that mostly you need to have people around. Population density is generally a necessity for public Wi-Fi access.

An important point is that Wi-Fi access requires infrastructure. There's a network "behind" Wi-Fi access, and the network provides a gateway to the Internet. Although it's theoretically possible to connect to the Internet via dial-up and then provide shared access via Wi-Fi, as a practical matter, most of the networks that provide Wi-Fi access use a broadband Internet connection. So to connect to the Internet anywhere without wires, you need the Wi-Fi radio broadcast, and you also need to have a solid connection to the Internet.

MOBILE COMPUTING ANYWHERE

A number of villages in northeastern Cambodia use Wi-Fi in an unusual way to connect with the Internet. Five men on mountain motorbikes connect these villages, which are otherwise too remote for Internet access, with the world. Each motorbike is equipped with a rugged portable computer equipped as a Wi-Fi access point.

Internet search queries and email are stored on the portable devices when each bike drives past solar-powered stations near the villages, which is linked to the villages using standard Ethernet cabling. Then the content is "dropped off," again using Wi-Fi technology, when the bike goes past a central satellite station that connects to the Internet.

The same process in reverse brings email (and answers to queries) back to the villages.

You can read more about this pioneering effort that uses Wi-Fi to bring the Internet to some of the world's most inaccessible places at http://www.firstmilesolutions.com/projects.htm.

Assuming that you don't live beyond the end of the Earth, where are the obvious places for Wi-Fi?

As I've already mentioned, it's probable that a Wi-Fi hotspot has a DSL or cable connection to the Internet. Most businesses have, or can get, this kind of access. In addition, for it to make sense for a business location to provide Wi-Fi access, it should be the kind of business in which one (or both) of two conditions apply:

- Revenue is generated when people decide they want to "hang out" for long periods of time.

- People are "stuck" in the place for long periods of time.

A third possibility, of course, is that providing wireless hotspot access is, in and of itself, a profit center.

Coffee shops are, of course, the canonical example of the first kind of Wi-Fi location (and sometimes, for example in the case of Starbucks, very much the third possibility as well). Airport waiting areas are probably the classical example of the second type of place that benefits from Wi-Fi, as in, "let me check my email while I'm waiting for my flight." Sadly, this is particularly true these days with the increased need to check in long before flights and the longer waits because of security concerns.

This chapter provides information about using available tools for finding Wi-Fi hotspots of all three types. The most useful tools are online directories, with, of course, the (sometimes big) drawback that you already have to be online to use them.

Of course, free hotspots—often put up by a vendor who hopes for increased business because of people hanging around using wireless networking, or choosing a particular business because it offers free wireless—are great. As one manager of a delicatessen restaurant told me recently, "I get much more business since I installed free Wi-Fi. My customers want to stay longer. And they do seem to get hungry as they surf the Web."

This chapter gives you some tips and techniques for finding free hotspots.

I'll also show you a neat gizmo—the Wi-Fi finder—you can use to see if there is a Wi-Fi network broadcasting nearby, without having to be online.

Next, I'll tell you what war driving and war chalking are about. These are two social movements that owe their origins to Wi-Fi and have to do with finding Wi-Fi networks. Practically, you are unlikely to ever have to find Wi-Fi hotspots using chalk marks on a sidewalk, but if you do see these marks at least you will know what they are. Having a look at these chalk marks is a good reminder of the culture of individualism, and making technology available free has helped to make wireless networking what it is today.

Finally, I'll show you how you might start to think of putting up your own hotspot, if this interests you.

Before I get started on this agenda, let me mention one simple, low-tech thing: It pays to ask. If you are looking for a Wi-Fi hotspot in an area far from home, just ask someone. Chances are that many people you meet can direct you to a local Wi-Fi hotspot—particularly if the person you ask is carrying a laptop with an Intel Centrino mobile technology logo.

Finding Hotspots

A *hotspot* is the term used to mean an area in which Wi-Fi users can connect to the Internet.

The term covers for-fee hotspots, which expect you to pay for access, just as you pay for Internet access via an Internet service provider (ISP), such as a cable or telephone company at home. It also covers free commercial hotspots, which are put up by businesses to generate traffic, as I've already discussed. Finally, more free hotspots are provided by institutions such as libraries and local governments. (For example, the city of Philadelphia is planning to provide hotspot access across virtually the entire city.)

In addition, there are "spill over" hotspots: put up with free access, but not necessarily your free access, in mind. Spill-over free access examples that are perfectly reasonable to use include hotspots put up by many business conventions, which provide Wi-Fi access as a courtesy to attendees, hotels providing courtesy Wi-Fi to guests in the lobby, and so on. However, you might want to think twice about "mooching" access from people who don't intend to share their Internet connection, but just don't have the savvy to protect it. (They probably haven't read this book.)

You might know in advance where to find Wi-Fi access on your travels. I've mentioned hotels and conventions already because these are likely places to find Wi-Fi access. You can certainly inquire ahead of time.

If you don't have advance information about the location of Wi-Fi hotspots, you can also just turn your laptop on and wander about from location to location like a digital Ulysses looking for wireless access. (You'll find a discussion of the Wi-Fi finder, a small gadget that might help you locate Wi-Fi hotspots, just by wandering, later in the chapter.) But assuming that you'd like something a little more pinpointed than the Clint Eastwood "Do you feel lucky?" approach, using the Internet to find Wi-Fi hotspots is the best way to go.

Of course, you have to be able to access the Internet directories from a location where you have Internet access. As Homer Simpson might say, "D'oh!" But this can become a catch-22, so you should certainly try to locate likely hotspots ahead of time (while you still have Internet access) if possible. I've seen people crawl from hotspot to hotspot, not leaving one until they've found the next.

There are three approaches to take when making your search for a place to surf:

- You can use the search tools provided by an organization whose branches host Wi-Fi hotspots, such as the Starbucks chain.

- If you've signed up with a Wi-Fi provider, you can search the directory of hotspots maintained by your service provider.

- You can search one of the many cross-provider Wi-Fi hotspot directories available on the Web.

I'll show how all three approaches might work using a test example. I live midway up in the hills in Berkeley, California. Let's suppose that I want to sip latte at a coffee shop and need to keep on checking my email while I do.

Searching a Chain

If you know the name of the organization or chain of stores that you would like to use as a wireless destination ("I want to surf at Starbucks," or "I want to browse at Borders"), you can go directly to the website of the organization to find a wireless location. The first approach, because I know that Starbucks coffee shops have Wi-Fi hotspots and I like Starbucks coffee just fine, is to find a Starbucks near me that is Wi-Fi enabled.

It's easy to go to `http://www.starbucks.com` and choose the store locator by clicking the Find Your Nearest Starbucks link on the home page. With the locator page open, I can select Wireless Hotspot Stores from the Store Type drop-down list, and fill in my city, state, and ZIP as shown in Figure 9.1.

Click Submit. You'll see a page showing the nearest Starbucks locations that are equipped with Wi-Fi along with a handy map (see Figure 9.2).

Starbucks is a likely chain to search for wireless access, but as more businesses decide it is worth their while to offer anywhere computing to their customers, there might be other choices that you are also interested in. For some other retail chains that feature Wi-Fi access, see Appendix B, "Where the Hotspots Are."

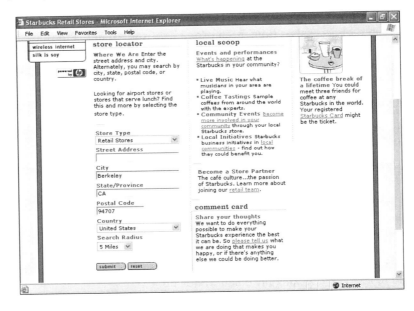

FIGURE 9.1

You can search on the Starbucks website for stores with Wi-Fi hotspots.

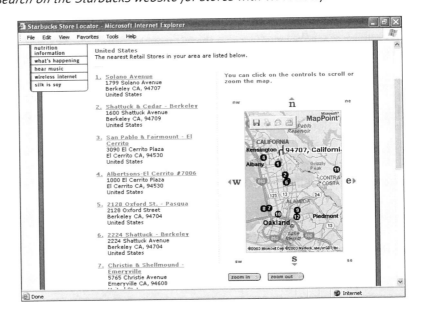

FIGURE 9.2

The Starbucks search shows a number of local stores with Wi-Fi hotspots.

Searching a Wi-Fi Service Provider

If you have signed up with a national network, such as T-Mobile Hotspot, it stands to reason that you will want to use hotspots provided by your network. For one thing, you probably have to pay a fee to your national provider. It's likely that you don't want to have to pay an additional fee to another provider. The best approach for finding network-specific hotspots is to use the directory provided by the network itself.

There are about a dozen major Wi-Fi service providers in the United States alone, and hundreds of smaller, mom-and-pop vendors. There are really only a couple of big Wi-Fi service providers with national coverage. If you've signed up for a payment plan with one of these big players, you will probably stick with the hotspots they provide so as not to pay multiple times for access.

Although there is no hard-and-fast rule about this, as with cell phone communications, you tend to get charged a bit more for "roaming." Because the industry is still so young and fragmented, there is, indeed, no guarantee that one Wi-Fi service provider has even set up a cross-billing arrangement to cover roaming with yours. As one Wi-Fi user told me, "I travel a lot, but I really don't want to get stuck paying two or more Wi-Fi providers a monthly service fee of at least $30 each."

Chapter 10 explains the structure of the Wi-Fi service provider industry, who the players are, how to pick the best one for you, and how to work with your Wi-Fi service provider. In the meantime, if you're searching for a national Wi-Fi service provider, three of the biggest are Boingo, Wayport, and T-Mobile Hotspot:

- Boingo Wireless has about 3,500 live hotspots in the United States, with a strong representation in hotels and coffee shops, and an international footprint. The Boingo website is `http://www.boingo.com`.

- Wayport is a privately held company based in Texas that is strong in hotels, airports, and—more recently—in McDonald's restaurants. The Wayport website is `http://www.wayport.com`.

- T-Mobile is a cell phone company that is a subsidiary of Deutsches Telekom. The T-Mobile Hotspot division provides Wi-Fi access in Borders, Starbucks, and many other locations—almost 5,000 nationally. The home page for T-Mobile Hotspot is located at `http://www.t-mobile.com/hotspot/`.

  T-Mobile provides a number of tools for searching for hotspots, such as the clickable map and drop-down list shown in Figure 9.3.

Ultimately, if I drill down on my location using the map T-Mobile provides, I'll get the same list of Starbucks locations provided by the Starbucks chain itself for Berkeley, California—not particularly surprising because Starbucks and T-Mobile are partners.

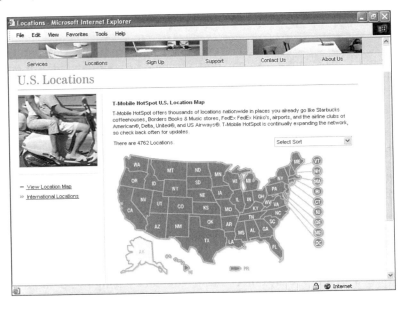

FIGURE 9.3

You can use the clickable map to find hotspots provided by T-Mobile in your state.

> **TIP**
>
> If you're just looking to get your feet wet in the pool of connecting to the Internet through a public hotspot, it's worth noting that Starbucks provides a promotional one-day pass that gives you a free day of Wi-Fi access using a special promotional code, as explained in special brochures available in each store. Other Wi-Fi vendors often provide comparable introductory offers.

Paying for Your Fun

As I've noted, there is an increasing movement on the part of retailers to provide free wireless networking—in part to drive traffic and in part as a matter of philosophy. I'm all in favor of this movement, as I really like getting my connectivity free.

However, it is still the case that a great deal of the time, as the saying goes, there is no such thing as a free lunch, or a free latte. For example, if you go to a Starbucks, you will need to pay to surf while you sip.

With most Wi-Fi service providers, you don't need to sign up for a plan, although it will certainly be beneficial for you to do so if you plan to make much use of the service. You also don't need to enroll with a Wi-Fi service provider in advance. You can do it on the spot when you connect via your laptop (provided, of course, you've brought a valid credit card).

With T-Mobile Hotspot, you can sign up for access by the minute, the day, or for a monthly plan.

The initial sign-up screen is shown in Figure 9.4 and gives you a good idea of the different plans and costs associated with them.

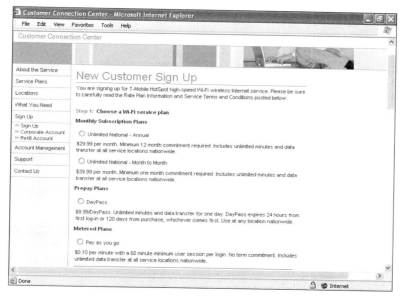

FIGURE 9.4

It's easy to sign up with T-Mobile or other Wi-Fi service providers.

The sign-up page shown in Figure 9.4, as you notice, is simply part of the T-Mobile Hotspot website, and once again you need to be connected to the Web to reach it. You'll be pleased to learn that you can also sign up from a wireless location without having done so in advance on the Web, as I'll show in a moment.

Surfing While You Sip

With your laptop equipped with Intel Centrino mobile technology, you are ready to surf while you sip. So what happens after you go to your local Starbucks?

First, make sure that the Starbucks provides wireless networking infrastructure (most do). If it provides a hotspot, it will have a T-Mobile Hotspot logo near the entrance (or you can ask one of the friendly Java slingers).

If you have never logged on to a T-Mobile Hotspot before, pick up one of the brochures that should be in the store. You can use the promotional code in the brochure to get free access for an initial day.

Fire up your computer. If the hotspot is up and running and there is good signal strength, Windows should detect the availability of a wireless network and tell you in an informational balloon message in the taskbar. (You can see this on the bottom right of Figure 9.5.)

FIGURE 9.5

A balloon message will alert you when wireless networks are present.

Double-clicking the wireless connection icon in the taskbar brings up the Wireless Network Connection dialog box, which prompts you to choose a wireless network. If you look at the list of available wireless networks you'll see that tmobile is there. This is good. Click on tmobile and the Connect button brings up the warning message (shown in Figure 9.6) that you are connecting to an unsecured network. Click Connect Anyway.

> This is a good place to remind you that public wireless networks are not secure by definition. You need to make sure that computers used at public hotspots are protected by turning off file sharing, turning on personal firewall protection, and taking other steps suggested in Chapter 17, "Protecting Your Mobile Wi-Fi Computer."

You should now be connected to the tmobile wireless network, as you can verify by glancing at the balloon above the connection icon on the Windows taskbar (see Figure 9.6). You can get wireless connection status information at any time by floating the cursor over the wireless connection icon on the taskbar.

Now would be a good time to check your Windows Firewall settings. You are in a public place, and who knows if there are people around you who might try to attack your computer through the network. As Captain Picard would say, "shields up!". Right-click the wireless connection icon in the taskbar and click the Change Windows Firewall Settings command, as shown in Figure 9.7.

The Windows Firewall properties window should appear, as shown in Figure 9.8.
Confirm that Windows Firewall is on.

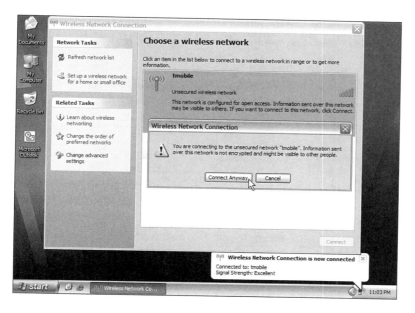

FIGURE 9.6

Windows generates a warning when you are about to connect to an unsecured network. A balloon message confirms that the connection is established.

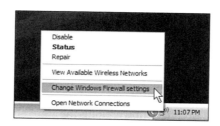

FIGURE 9.7

You can conveniently access Windows Firewall settings from the wireless connection icon in the system tray.

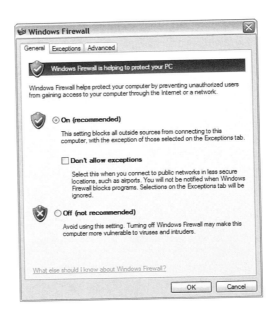

FIGURE 9.8

The radio button next to On (recommended) should be selected. If it's not, click it to turn Windows Firewall on.

> **TIP**
>
> If you are using a browser other than Internet Explorer—for example, Mozilla or Firefox, you might have problems accessing the login page. In this case, you should switch to Explorer to log in, and after logging in you can return to using the browser of your choice.

There's one more step to take before you can surf the Internet and check your mail—you must arrange to pay for your fun. To do this, open a Web browser such as Internet Explorer. It doesn't matter what you have set your browser's home page to, you will be automatically redirected to the Starbucks T-Mobile Hotspot sign-on page shown in Figure 9.9.

If you already have a T-Mobile Hotspot account, you can enter your username and password in the upper right and click Go to use the Internet in a normal fashion. If you need to establish an account, click the Sign Up link (shown along the right side of Figure 9.9). The normal T-Mobile account sign up page, shown back in Figure 9.4, will open, and you can choose the plan that best suits you. Be sure to enter the promotional code for free initial access if you choose to take advantage of this.

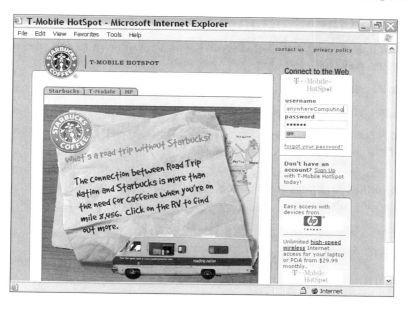

FIGURE 9.9

The first time you open your browser, you are re-directed to a page for logging on to the T-Mobile Hotspot network.

Using Directories

The best and easiest way to find Wi-Fi hotspots is to use the directories available on the Web. There are only two problems with using these directories:

- As I've already mentioned, it can be a significant problem that you have to be already connected to the Internet to access them.

- No one directory is comprehensive, and the information in each directory is different. This means that you might need to search multiple directories to find the hotspot you are looking for.

Where to Find the Directories

Here are some of the most widely used Wi-Fi hotspot directories and where to find them on the Web:

- China Pulse: `http://www.chinapulse.com/wifi` (hotspot locater for China).

- HotSpotList: `http://www.wi-fihotspotlist.com`.

- i-Spot Access: `http://www.i-spotaccess.com/directory.asp` (currently limited to Illinois, Iowa, Missouri, and Nebraska).

- Intel's Hotspot Finder: `http://intel.jiwire.com`. Features hotspots that have been verified for Intel Centrino mobile technology. Hotspot information also appears on relevant Yahoo maps.

- JIWIRE: `http://www.jiwire.com`.

- Ordnance Survey: `http://www.ordnancesurvey.co.uk/oswebsite/business/sectors/wireless/wifihotspot.html` (great for United Kingdom hotspots).

- Square 7: `http://www.square7.com/btopenzone/directory.htm` (great for European hotspots) .

- WiFi411: `http://www.wifi411.com`.

- Wi-Fi-Freespot Directory: `http://www.wififreespot.com`.

- WiFinder: `http://www.wifinder.com` (one of the best all-around international hotspot directories).

- WiFiMaps: `http://www.wifimaps.com`.

- Wireless Access List: `http://www.ezgoal.com/hotspots/` (categorized by state and ZIP Code, also allows sorting by network—for example, T-Mobile, Wayport, and so on).

Which Directory to Use

As I mentioned earlier, if you are looking for a Wi-Fi hotspot provided by a specific network provider, that network provider is the best source of location information.

If you need to find a hotspot in a specific geographic area, a directory that targets a specific area might be the way to go. For example, China Pulse has the best listings for finding hotspots in China (yes, Virginia, there is Wi-Fi coverage in China), and the Ordnance Survey is one of the best way to find Wi-Fi hotspots in Great Britain.

Otherwise, I find that the best all-around directories for use with commercial Wi-Fi hotspots are HotSpotList and WiFi411.

Getting Around the "You Have to Be Connected" Barrier

The JIWIRE directory, `http://www.jiwire.com`, has a partial answer to the "you can't get there from here" problem.

JIWIRE has packaged its hotspot directory for easy access by multiple platforms. Supported platforms include Windows, Mac OS, AvantGo handhelds, and online access from cell phones with web browsers. See `http://www.jiwire.com/hotspot-locator-frontdoor.htm` for more information.

The downloads for Windows and Mac OS are applications that contain the JIWIRE hotspot location database. To keep up to date the application periodically connects to the Internet to get revisions.

To use the handheld directory, you need an AvantGo account (`http://www.avantgo.com`) and a wireless handheld capable of working with the AvantGo network. You download the AvantGo Hotspot Locator (powered by JIWIRE) using your desktop computer following the instructions on the JIWIRE and AvantGo sites. The next time you synch your handheld with your desktop, the AvantGo Hotspot Locator will be downloaded on to the handheld. You will be able to use it much as you use a desktop Wi-Fi directory.

To use the WAP-enabled phone directory, you need a wireless phone with a WAP browser and a mobile Web account with your cell phone provider. You can then use the WAP-enabled phone as a Wi-Fi finder by using it to surf to `http://wap.jiwire.com`. Of course, depending on your service plan, you can expect to be charged for the time you are searching.

Tricks and Techniques

If you've spent much time surfing the Web, you've probably learned some techniques for dealing with search engines provided by sites such as Google, Yahoo!, and even Amazon. Searching a Wi-Fi directory for a hotspot is much the same: There are different tricks and techniques for using each one. In this section, I'll take a look at using my two favorite Wi-Fi directories, HotSpotList and WiFi411.

Using the HotSpotList Directory

Figure 9.10 shows the HotSpotList main window.

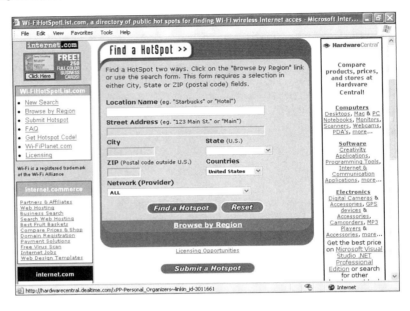

FIGURE 9.10

The HotSpotList search window.

This window is used to conduct searches. (You can also browse by location, as I'll show you in a second.)

To conduct a search, you need to enter a business name or type, and some geographic information (such as city and state, or a ZIP Code). For example, a search for Starbucks in New York provides many pages of results, as you can see in Figure 9.11.

> **NOTE**
>
> The HotSpotList network provider filter includes all the Wi-Fi networks you might ever have heard of, such as those listed in Appendix B. It also includes many "boutique" networks that you have probably never heard of.

A nice feature of this search window is that it allows you to screen by network provider. When you open the search window, it is set to show all network providers, meaning that there is no network provider screening. However, you can select a network provider, such as Boingo Wireless, from the drop-down list, shown in part in Figure 9.12.

If you go back to the HotSpotList main search window and click the Browse by Region link, you can then choose a country to browse from the list.

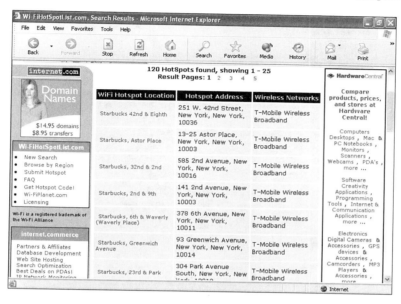

FIGURE 9.11

The search results window.

Most of the countries shown in the list are not further subdivided, but if you click on the United States link, you can then click a link for each state.

Drilling down the next step, if you click on a state link, you will see the cities within that state with hotspots (see Figure 9.13).

If you click on a link representing a city, you will then see the listings for that city. For example, Figure 9.14 shows the listings for my hometown, Berkeley, California.

The detailed listing provides the address of the hotspot, the Wi-Fi network provider, and a link to more detailed information about the hotspot. For example, Figure 9.15 shows the listing for Pacific Ironworks, a local climbing gym where my son Julian had his sixth birthday party that offers free Wi-Fi access. (So I could surf while he climbed.)

If you ever plan to visit beautiful, downtown Berkeley, be sure to bring your mobile Wi-Fi device and visit Pacific Ironworks. Even if you don't do any climbing, you can certainly surf the Web.

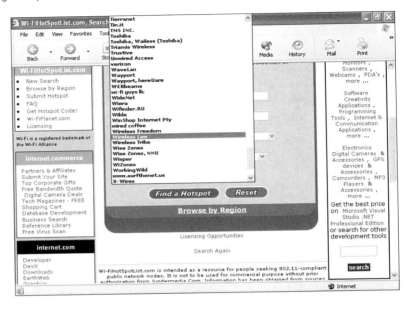

FIGURE 9.12

You can filter the HotSpotList search by network provider.

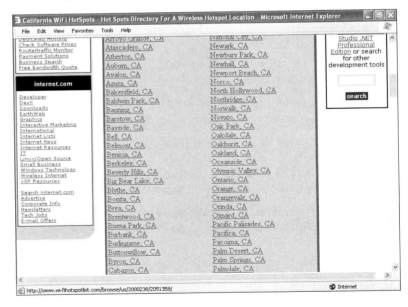

FIGURE 9.13

Each city within the state that has at least one hotspot is shown.

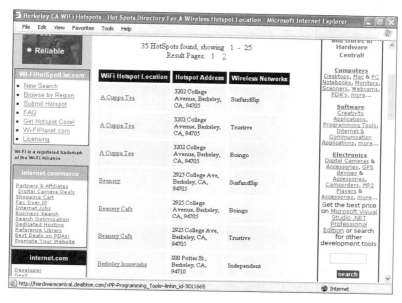

FIGURE 9.14

The HotSpotList listings for Berkeley, California.

Using the WiFi411 Directory

The main window of WiFi411, shown in Figure 9.16, doesn't provide a way to drill down by region (as HotSpotList does), but it does add a few useful bells and whistles of its own.

One of the most useful things about WiFi411 is that you can search by type of hotspot. Figure 9.17 shows some of the results of searching for a coffee shop in Berkeley.

The type of search is determined by selecting a type from the drop-down list shown in Figure 9.18. (You can also choose All types, and not filter based on type.)

Perhaps the most useful feature of the site is the roaming feature. If you filter your search results using a Wi-Fi network provider, when the results are returned, a Roaming check box is displayed. (You can see this check box in Figure 9.19.)

If you check the Roaming box and search again, the new results will show not only your network provider's hotspots, but also hotspots that are served by any network with whom your provider has a roaming arrangement.

You might not want to roam; you might just want to stay home. But whatever your call, WiFi411 is one of the best hotspot locators of all.

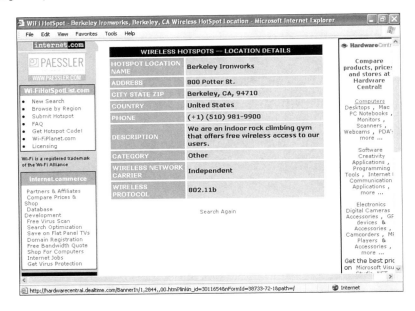

FIGURE 9.15

Detailed listings provide further information about each hotspot.

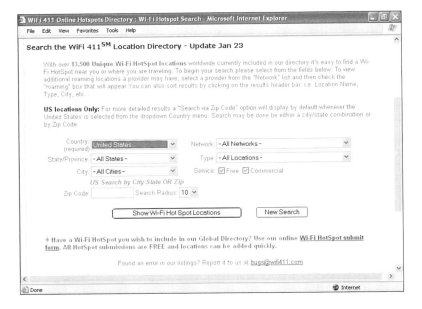

FIGURE 9.16

The WiFi411 main window.

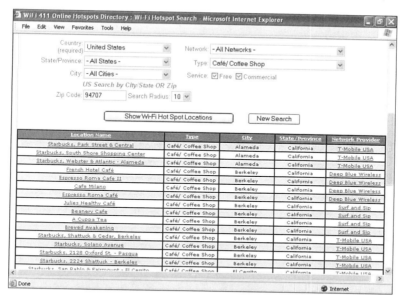

FIGURE 9.17

Search results for a coffee shop in Berkeley.

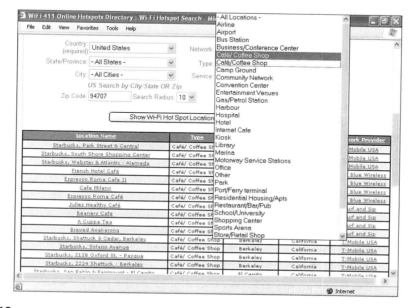

FIGURE 9.18

You can search for a particular type of hotspot by searching from the list.

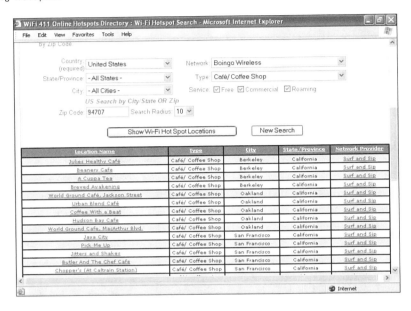

FIGURE 9.19

The Roaming check box appears when you select a network provider.

Finding Free Hotspots

I like getting my wireless networking access without having to pay for it—and I'll bet you do, too.

The good news is that free access is spreading like wildfire among commercial establishments. (I'm not talking about mooching wireless service from a neighbor with an unprotected access point.)

This movement started when independent businesses, such as unaffiliated coffee shops, discovered that by offering free access they could successfully compete with big chains, such as Starbucks, where access was on a fee basis. Offering free access is also in keeping with the original spirit of wireless networking and the Internet, which is to make things as widely available to everyone as possible.

Quite a number of business chains have also decided that providing free access makes business sense. (Some of these chains are listed in Appendix B.)

For now, free wireless access models exists alongside the pay-for-wireless service model. We'll have to see which one (or both) survive in the long run.

You can vote with your feet and patronize establishments that provide free access.

One of the best ways to find free sites is to use the Wi-Fi-Freespot directory, `http://www.wififreespot.com`, shown in Figure 9.20, which provides a state-by-state listing of free hotspots.

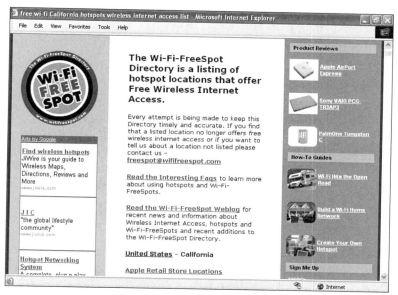

FIGURE 9.20

The Wi-Fi-Freespot directory provides a state-by-state listing of free wireless hotspots.

Some of the other directories—such as WiFi411, which I described earlier in this chapter—allow you to check a box that filters results so that only free hotspots are shown. So be aware that many of the search directories allow you to distinguish between free and for-pay wireless hotspots. You might want to be sure to use one of these if your credo is "I've paid and I've done it free, and I won't go back to paying anymore." The answer to the question of what is the difference between free wireless access and access that costs $30 a month is likely to be $30 a month.

I've already mentioned that quite a few retail chains have taken to providing free access as a matter of policy (see Appendix B). You should also know that many local libraries offer free wireless access (check with your local public library). Finally, another interesting possibility is a college or university. These days, most campuses have a wireless infrastructure. Even if it is only intended for the use of students and faculty, it might be possible to arrange for guest access.

To locate free hotspots, don't forget word of mouth, asking people, observation (the point of war chalking, which I'll describe in a little while), and searching for signals using your mobile computer or a Wi-Fi finder.

Working with a Wi-Fi Finder

One approach to finding Wi-Fi hotspots is to use a Wi-Fi finder, a small, inexpensive device that has appeared on the market recently.

The most commonly used Wi-Fi finder is the one from Kensington Technology Group, model number 33063. This tiny unit measures 2.7×3 inches, and is about 1/2 inch thick.

It's simple to use the Wi-Fi finder. When you're out wandering, press the unit's button. If it detects an 802.11b or 802.11g Wi-Fi network, the green lights illuminate. The more green lights, the stronger the signal. If only the red light illuminates, there's no Wi-Fi network in the neighborhood.

> **NOTE**
>
> A Wi-Fi hotspot is termed *open* when it is not encrypted. This means that you do not have to supply an encryption key, which looks and acts like a password. An example of an open hotspot is the T-Mobile Hotspot hosts, which require an SSID but not an encryption key. (You also have to log on to the T-Mobile network with your T-Mobile user ID and password, but that's a different issue.)
>
> A Wi-Fi hotspot is *closed* when an encryption key is required to access the node. All private Wi-Fi networks should be run in closed mode.
>
> Wi-Fi encryption is accomplished using WEP (Wireless Equivalency Protocol), a security measure that is part of the 802.11 and Wi-Fi standards.

The Wi-Fi finder blocks out competing 2.4GHz signals from devices such as cordless phones and microwaves, so that when it shows a signal, you know it is from a Wi-Fi network. It has an effective range of about 200 feet out-of-doors.

So far, so good. But the fact of the matter is that you can't tell from the Wi-Fi finder what the SSID of the network is, who the network provider is, whether the hotspot is free or commercial, whether the network is open or closed, and if closed, who to contact for access.

So a Wi-Fi finder is a fun and pretty cool thing to put on your keychain. (The Kensington model comes with a ring for attachment to a keychain.) But is it genuinely useful? That probably depends on the user, but you can definitely find me out cruising the neighborhood to see where the Wi-Fi is using my handy-dandy Wi-Fi finder (see Figure 9.21).

War Driving and War Chalking

War driving is the hobby of popping in a car and cruising around with a Wi-Fi–equipped laptop looking for open Wi-Fi nodes.

War chalking is the act of using specific chalk markings, usually on a sidewalk, to identify Wi-Fi hotspots.

Both activities show the extent to which Wi-Fi has come out of nothing as a disruptive technology, until fairly recently because of grassroots support and without the backing of major corporations. Although important companies such as Intel and others now back Wi-Fi, the "counterculture" roots of old-time Wi-Fi and 802.11 practitioners remain an important aspect of the social history of Wi-Fi.

FIGURE 9.21

The author is cruising for Wi-Fi in all the wrong places.

War driving is an activity that ranges from the fun, but harmless (that is, if you find this kind of thing fun) to malicious and felonious (theft of information such as credit card numbers and industrial spying). It's successful only because an astoundingly large number of corporations and others that deploy private Wi-Fi networks do not use even rudimentary protection such as encryption. It has been estimated that as many as 60% of the privately deployed Wi-Fi networks operate without encryption.

Supposedly, war chalking is derived in part from the marks hobos made during the Great Depression. These hobo chalk markings would point out a place to get a free meal, a good place to sleep, and perhaps a particularly nasty railroad "bull."

In war chalking, an open node (public hotspot) is indicated by two opposing half circles marked with the SSID and the bandwidth (see Figure 9.22).

FIGURE 9.22

The war chalk marking for an open node.

A closed node (or private Wi-Fi network) is shown as a circle, along with its SSID, as in Figure 9.23.

FIGURE 9.23

The war chalk marking for a closed node is a circle.

A Wi-Fi node protected by WEP encryption is shown as a circle with a letter W within it (see Figure 9.24). Ideally, the encrypted node is marked with SSID, bandwidth, and access contact. (The access contact can provide the encryption key if it is appropriate.)

FIGURE 9.24

The war chalk marking for a WEP node.

The fact that war chalking has spread as a practice is a testament to the power of the Internet to spread social habits. It also speaks to the problem of finding Wi-Fi hotspots if you are not already online. In the future, it's likely that Wi-Fi hotspot information will become more ubiquitous. Perhaps the community welcome signs that now tell the time and place for Rotary Club meetings will soon come to also provide Wi-Fi hotspot information. When travelers visiting a community can find Wi-Fi hotspot information in this kind of way, war chalking will probably fade into history like the chalk marks made by the hobos of long ago.

WHERE DOES THE "WAR" IN "WAR DRIVING" AND "WAR CHALKING" COME FROM?

Why does "war driving" use the term "war?" Admittedly, the practice of war driving is an attempt to obtain information illicitly, but a better term might be something like "eavesdrop driving."

The term "war chalking" seems to have come along as a Johnny-come-lately term for "war driving." War driving was already firmly entrenched in the Wi-Fi counterculture, so when the newer grassroots activity of marking Wi-Fi hotspots came along, it seemed natural to use the "war" word. But how was "war driving" named in the first place?

The term derives from "war dialing," which was the pre-broadband hacker practice of programming a computer to dial hundreds of dial-up access numbers, hoping to find one that was not protected (or could easily be cracked). It was popularized by the movie *WarGames* (1983), starring Mathew Broderick.

Becoming a Hotspot

Well, I don't think that you really want to *be* a hotspot—but perhaps you might want to put one up so that others could use it.

If you run any kind of small business, this might make a great deal of sense. By way of comparison, Schlotzsky's, Inc., which runs deli restaurants, has stated that adding free Wi-Fi to its shops adds more than $100,000 revenue for each store per year (through added purchasing by customers who come to the store for the Wi-Fi hotspot or who stay longer than they otherwise would).

You might also want to put up a Wi-Fi hotspot simply as a service to your fellow humans. (Believe it or not, this kind of altruism has largely sparked the growth of Wi-Fi.)

The technical aspects of putting up a Wi-Fi hotspot, meaning the hardware infrastructure required, don't differ that much from putting up a Wi-Fi network for personal or business use. To start with, you need a broadband connection. If you are planning to resell access via a Wi-Fi hotspot, most cable and DSL providers will require you to buy a commercial-grade account (rather than a personal use account). It's also the case that you should probably expect to use commercial-grade equipment (rather than that intended for homes and small offices) if you expect any traffic at all.

The problem is not the hardware so much as what is called, in the telecommunications business, *provisioning*. Provisioning means setting up the systems that provide customer service, support, and billing.

You'll want to consider provisioning issues even if you plan to give away free Wi-Fi access because it is dangerous to allow unrestricted access to your network. If access to your network does not require registration, it could be used for malicious purposes—for example, spamming, which could get your Internet address blacklisted.

Some of the national Wi-Fi networks will provide basic support and protection for IP address abuse for a reasonable fee. If you are interested in this, you should check out Surf and Sip, http://www.surfandsip.com, which specializes in Wi-Fi–enabled hotspots for providers who want to give away free access.

If selling access is more your cup of tea (or Java), there are any number of companies that sell turnkey packages. Generally, you purchase the hardware from the company, which then provides provisioning services and splits the proceeds from billing using an agreed-upon percentage.

> **NOTE** Hosting a Wi-Fi hotspot can lead to serious security risks. Even if you are not charging for access, at the very least, you should probably consider instituting a user authentication scheme. If you choose to provide an unprotected network for others to use, even if they are only your neighbors, be aware that you might have some liability for illegal actions—such as denial of service attacks—that use your wireless network.

One turnkey provider of this sort is Pacific Wi-Fi, http://www.pacificwi-fi.com. Another similar product is Instant Hotspot from Advanced Internet Access (see http://www.instanthotspot.com/wsg5000.htm for more information) .

When you have your Wi-Fi hotspot up, your problem is the reverse of the one primarily discussed in this chapter. You don't need to find a hotspot—you need people to find your hotspot. If you don't get the word out, no one will know about it. A hotspot that has not been promoted has a place in the world like the proverbial tree that falls in the forest—if no one knows it has fallen, what is the point? (Or is it even real?)

An essential first step in promoting your new hotspot is to make sure that it appears in the directories described earlier in this chapter.

Of course, after you put up your hotspot you could "war chalk" it yourself using the standard symbols explained in this chapter. (Traditionally, these symbols are chalked on the sidewalk, but they could also go on a permanent sign.) In a heavily trafficked area, a simple sign that says "Wi-Fi Hotspot" would probably also draw traffic.

Summary

Here are the key points to remember from this chapter:

- There is no one single good way to locate all Wi-Fi hotspots.

- It's easy to "get down and boogie" using wireless at Starbucks or other locations hosting hotspots.

- Online directories are a great source of information about Wi-Fi hotspots and are central to the Wi-Fi movement.

- It's problematic finding Wi-Fi hotspot information if you are not already online. Solutions range from asking people, to using a Wi-Fi finder, to reading war chalk marks.

- Besides being a wireless protocol, Wi-Fi is a way of life and a cultural movement.

Working with National Wi-Fi Networks

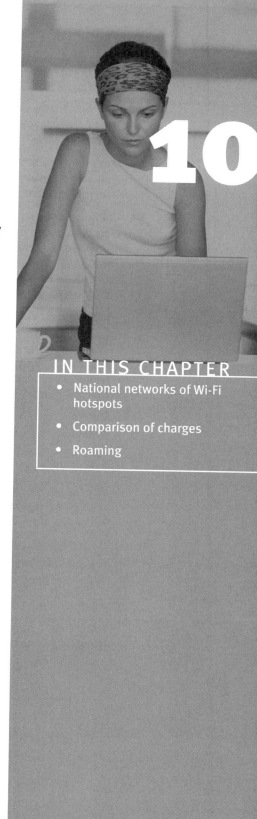

In this chapter, I suppose that you have your laptop equipped with Intel Centrino mobile technology ready to roll and want to sign up with a Wi-Fi network. You might primarily be interested in using Wi-Fi hotspots locally, or you might want to use them for more distant traveling. In either case, the issues are much the same as when you choose a wireless cell phone provider. The benefit of having a national provider is that you can login from multiple locations and around the country without having to separately sign up (and pay) each time. You should be concerned about the following:

- How widespread is the coverage?

- What are the costs and fees?

- Are there any discounts?

- What are the roaming policies in place?

- How good is customer service?

This chapter answers these questions. I'll start with a run-down of the major national networks. Depending on your needs and location, you might be interested in the more complete listing of Wi-Fi networks with some of the smaller players and some non–U.S. providers in Appendix B, "Where the Hotspots Are."

Wi-Fi Networks

National Wi-Fi networks have been emerging from the top down and from the bottom up. Top-down network vendors—such as AT&T Wireless, T-Mobile Hotspot, and Verizon—were already in the telecommunications business. These companies decided that there was a business opportunity in meeting the needs of people like you and me by creating a Wi-Fi network.

The bottom-up vendors, such as Boingo Wireless and Wayport, came from a different place. They were entrepreneurs who figured that there was an opportunity in the new, "disruptive" Wi-Fi technology.

Although the business of providing Wi-Fi access remains fragmented, some major national networks are coalescing. It has become clear that to survive from a business viewpoint, the networks need to reach a critical mass.

Table 10.1 shows the national U.S. networks that seem to be approaching the size of infrastructure (meaning the number of hotspots) necessary for long-term survival. I've provided telephone numbers and Web addresses for the networks. This contact information might be useful if you want to do further research or if you decide that you want to sign up with a specific network. It could also help if you are traveling to an area that is particularly well served by a specific provider.

Table 10.1   Wi-Fi Networks

Network	URL	Phone	Comments
Cingular Wireless	`http://www.cingular.com sbusiness/wifi/`	888-290-4613	Cingular's Wi-Fi service is formerly AT&T.
Boingo Wireless	`http://www.boingo.com`	800-880-4117	A pioneer Wi-Fi network, and still considered one of the best. Great roaming policies.
Sprint PCS	`http://www.sprint.com/ pcsbusiness/products_ services/data/wifi/ index.html` or `http://www.wifi.sprintpcs.com`	866-727-9434	Rolling out quickly based on existing infrastructure.
T-Mobile Hotspot	`http://www.t-mobile.com/ hotspot/`	877-822-7768	The leading Wi-Fi provider, building infrastructure in Starbucks stores and elsewhere. T-Mobile is the one everyone else wants to beat.

Network	URL	Phone	Comments
Wayport	`http://www.wayport.com`	888-929-7678	An early pioneer in providing Wi-Fi hotspots, now providing infrastructure in McDonald's restaurants and elsewhere.

Special Pricing, Good and Bad

You should be aware that there are likely to be all kinds of special pricing deals when you sign up for Wi-Fi access. Mostly, this is all to the good.

> **TIP** The Wi-Fi industry is rapidly changing. Be sure to verify that the information in this chapter is current when doing your own research.

For example, most establishments that provide fee-based Wi-Fi access also have some special, introductory offers. In this spirit, it is typical to find a coupon at Starbucks good for a one-day pass on T-Mobile Hotspot.

However, you should also know that some Wi-Fi networks, such as Wayport, allow individual hotspot operators who are part of their network to charge more than the standard network price for access. If the location charges more than the standard network fee, your credit card will be billed for the overage, so buyer beware. If there's any doubt, be sure that you verify prices for access at a particular location.

What the Networks Charge

The charges and payment structure in this section are based on information published by each network or on conversations I have had with Wi-Fi network company representatives. As I've noted before, they are certainly subject to change, so you should verify current pricing for yourself. I've tried to include some tips and ideas for getting the most out of each Wi-Fi network.

Cingular Wireless

Cingular Wi-Fi service, called Laptop Connect, offers a number of different Wi-Fi service plans, as shown in Table 10.2.

Table 10.2   Cingular Wireless Wi-Fi Plans

Plan	Details	Price
1 Time Connect	One-time 24 hour access at the location at which access was purchased.	$9.99
5 Connect Package	Five one-time connects; valid for six months (note that each connect is only good at a single location).	$29.99
10 Connect Package	Ten one-time connects; valid for six months (note that each connect is only good at a single location).	$49.99
Monthly Unlimited	"Unlimited" usage at unlimited locations. Note that this "unlimited" plan is actually limited to 150 connections per month (for "antifraud" reasons). Although 150 connections per month is a lot, one could easily see how it might be exceeded.	$69.99 per month

Cingular Wireless has a substantial presence in Amtrak train stations, airports, and—via its roaming arrangements with Wayport and Stayonline—more than 500 hotels.

Boingo Wireless

Boingo Wireless has two kinds of payment plans available as shown in Table 10.3.

Table 10.3   Boingo Wireless Wi-Fi Plans

Plan	Details	Price
Boingo AsYouGo	Unlimited daily access from any number of locations	$7.95 for the first two days; $7.95 for each additional day thereafter
Boingo Unlimited	True unlimited monthly access from any number of locations	$21.95 per month for the first year; $31.95 per month thereafter

Boingo Wireless has more than 3,500 locations, including 112 hotspots at airports, 1,175 hotspots at restaurants and coffee shops, 1,429 hotspots at hotels, and 19 hotspots at convention centers. In addition, Boingo has extensive roaming arrangements with Wayport, Surf and Sip, and many smaller networks. If you download Boingo's special software, you can use roaming networks seamlessly. (You won't even know that you are roaming).

Otherwise, if you connect normally to a hotspot, you can enter your Boingo user ID and password to log on to the "foreign" network. The ability to effectively roam on other networks greatly extends Boingo's Wi-Fi network.

Sprint PCS

Sprint PCS offers Wi-Fi access on a daily basis or as part of its business data plans as shown in Table 10.4. The Sprint business data plans shown in Table 10.4 are primarily intended for business customers. Costs are based on the amount of data transferred between your mobile computer and the Wi-Fi network.

> **TIP**
> In some cases, you can save money by using Boingo for access and then roaming on a foreign (and nominally more costly) Wi-Fi network.

Table 10.4   Sprint PCS Wi-Fi Plans

Plan	Details	Price
Pay-As-You-Go	Unlimited daily access from a particular location	$9.95
Month-to-Month	Unlimited monthly access	$49.95 per month

Sprint PCS has no Wi-Fi roaming arrangements available, but is rapidly rolling out hotspots across its network, particularly in airports and hotels.

T-Mobile Hotspot

T-Mobile Hotspot is the leading provider of Wi-Fi hotspot services and perhaps the old-line telecommunications company that has successfully created a true national Wi-Fi network. (T-Mobile is a subsidiary of Deutsche Telekom).

This effort is spearheaded, of course, by the presence of T-Mobile Wi-Fi access points in Starbucks shops. There are also T-Mobile Hotspots in most Borders bookstores and in many airports and hotels.

Table 10.5 shows the T-Mobile Hotspot pricing plans.

Table 10.5   T-Mobile Hotspot Wi-Fi Plans

Plan	Details	Price
Metered Plan	This is a no-strings-attached, pay-as-you-go plan. No term commitment, valid at all locations.	$0.10 per minute, with a 60-minute minimum
DayPass	Unlimited access from all locations for twenty-four hours.	$9.99
Unlimited National—Monthly	Unlimited monthly access.	$39.99 per month
Unlimited National—Annual	Unlimited monthly access with a price break when you commit to a year.	$29.99 per month

T-Mobile offers no roaming facilities on other networks.

Wayport

Wayport is a Wi-Fi pioneer, with massive presences particularly in hotels and airports.

The Wayport pricing plans are shown in Table 10.6. (Because Wayport allows vendors who put up one of their hotspots to markup their single-use fees, you should verify costs.)

Table 10.6   Wayport Wi-Fi Plans

Plan	Details	Price
Single hotel connection	Unlimited access from a guest room or common area. Note that guest rooms and common areas are considered to be separate locations, and charged separately. (You incur two fees.)	$9.95 until next hotel check-in time. (Hotel might mark this up, so you should check because it might vary.)
Single airport connection	Unlimited access from an airport.	$6.95 from purchase time through midnight.

Plan	Details	Price
Prepaid connection cards	Wayport connection cards give you time-limited access. (The amount of access varies by location.) These cards can be bought at Wayport's Laptop Lane facilities, found in many airports. (You'll also get a Laptop Lane discount when you buy one.)	$25 for 3 connections. $50 for 8 connections. $100 for 20 connections.
Month-to-month membership	Unlimited monthly access.	$49.95 per month.
Annual membership	Unlimited monthly access with a price break when you commit to a year.	$29.95 per month with a one-year contract.

Wayport does not offer roaming on other networks to its customers.

Comparison Shopping

Don't be thrown by the complexities of all the different Wi-Fi pricing models. It's not really as complicated as it might seem.

As a practical matter, the first time you use a Wi-Fi hotspot, you'll probably take advantage of a promotional offer, or buy one-time (or pay as you go) access. It's a good idea to stay uncommitted for a while and to try a variety of different networks.

So start with one-off usage, and get a feeling for a number of the Wi-Fi hotspot networks. You should make note of locations, access speeds, and how good the customer service is.

After you've used a number of Wi-Fi hotspots, you should begin to get the sense of your usage patterns, and you might be ready to sign up for an extended payment plan by the month (or even an annual contract). If you keep a log showing your actual usage and compare it to pricing explained in this section, you might be able to come up with the best pricing comparison.

If a particular location is vital to you, you might be stuck with one particular provider, and comparisons won't matter so much.

The roaming features offered by Boingo and (to some degree) Cingular Wireless make these networks possibly more attractive than the others.

It's hard to beat the massive coverage of T-Mobile Hotspot with all those Starbucks locations. In other words, if you can afford the T-Mobile fees, they are probably the best choice going. However, if you are primarily a business traveler, this might not help you as much as availability in airports and hotels.

From a customer service viewpoint, you might be happier with one of the networks whose primary business is Wi-Fi—Boingo, T-Mobile, and Wayport—as opposed to one of the old-line companies for whom it is an afterthought (meaning Cingular Wireless and Sprint).

Summary

Here are the key points to remember from this chapter:

- A limited number of national Wi-Fi networks provide hotspots.
- A few of these networks (Boingo, Wayport) are attempting to build Wi-Fi networks from scratch.
- Other networks use the infrastructure of wireless or wireless telecommunications providers. (T-Mobile Hotspot is the leader of these.)
- Boingo is the network with the best roaming policy.
- Pricing is erratic and varies from network to network.
- You can generally purchase one-time access, access by the day, or monthly access.
- Some Wi-Fi network providers allow marked-up pricing; beware of retailers who overcharge for Wi-Fi access.

YOUR OWN WIRELESS NETWORK

PART IV

Networking Without Wires

Wireless networking—Wi-Fi—is a great tool for more than mobile computing. You can use Wi-Fi at home or in the office to construct a network without the hassle (and expense) of dealing with wires and drilling holes through the walls. Furthermore, using Wi-Fi for your home or small office network gives you the freedom to work wherever you'd like: in the garden, in the kitchen, in bed, or maybe even in a bubble bath. (Where are waterproof laptops when you need them?)

This chapter provides an overview of what it takes to set up a wireless network in the wondrous world that has no snaking tangles of wire. I'll also point you in the right direction for more in-depth coverage in this book of specific topics related to wireless networking.

Understanding Home and SOHO Networks

If you are used to working on a single computer, the idea of setting up a network might seem daunting. Perhaps at work you *do* plug in to the corporate network; however, maintaining and configuring this network isn't your problem but is instead handled by a staff of highly professional overachievers. At least, that's what the folks from corporate information technology (IT) would have you believe.

Relax! There's nothing particularly dark, deep, or mysterious about the concepts involved in setting up a small home or office network.

I'd like to step back for a moment or two and forget about Wi-Fi and wireless connectivity. This will give me the chance to explain networks to you generally. As you'll see, networks are really simple. There are no really tough concepts involved. By explaining the concepts and showing you the relevant vocabulary, I can help make sure that you'll make the right purchasing decisions (and never be snowed by a salesperson's jargon).

11

The only real difference between wired and wireless networks is the following:

Wired   Information is sent and received using the wire connections.

Wireless Radio transmissions are used.

In the Beginning, There Was the Connection

At its most basic level, a network is simply two or more computers or devices that are connected, as shown in Figure 11.1.

> **NOTE**
> Many devices can be added to a network. A good example is a network-enabled printer. However, in this chapter, I'm pretty much just going to talk about computers and devices generically, and I'll essentially be using the terms "computer" and "device" synonymously.

Most modern networks, including the Internet itself, use a protocol called TCP/IP (Transmission Control Protocol/Internet Protocol) to standardize communications.

You don't really need to know anything about TCP/IP to use wireless networking (or wired networking, for that matter). But you might be interested to know that TCP/IP consists of a number of different so-called *layers* that specify how network transmissions are broken down into units, called *packets*, and reassembled, and much more.

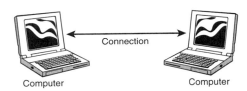

FIGURE 11.1

The simplest network consists of two connected computers.

TCP/IP is distinct from the mechanism used to convey the communication, meaning that if your network is operating over a wired connection, such as 10BASE-T Ethernet, the TCP/IP transmissions pass "over" the 10BASE-T wires. Similarly, if your network operates using Wi-Fi, TCP/IP transmissions are occurring "on top of" the Wi-Fi signals.

To Serve or Not to Serve

From a practical viewpoint, there are really two different ways that a network can be arranged. The arrangement of a network is called a network *topology*.

The simplest setup is one in which computers share resources such as files, printers, and Internet access on an ad-hoc basis. This is often called a *peer-to-peer* network. At a concept level, which means forgetting about things such as whether the connections are made with wires or radio waves, a peer-to-peer network might resemble the one shown in Figure 11.2.

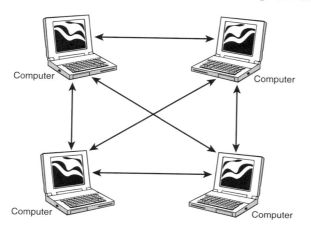

FIGURE 11.2

In a peer-to-peer network, computers share their resources.

The other type of network topology, client/server, is somewhat more complex. In this kind of setup, a centralized server computer controls and polices many of the basic functions of the network. For example, the server is used to authenticate users and to make sure that they have permission to take specific actions in respect to resources. In this kind of setup, only specific users (or kinds of users) may be allowed to modify or delete files. (Although individual computers can share resources directly, the sharing can only take place if the policies established on the server allow it.)

It's hard to enforce this kind of policy on a network without a centralized server. At a conceptual level, forgetting for the moment how the computers are actually connected, a client/server network might resemble the one shown in Figure 11.3.

Generally, client/server networks are found in larger-scale enterprise environments. I've described them here in case you work in an environment that has this kind of network. But you probably don't need anything as complex (and expensive) to administer in your home or small office. For the sake of keeping things simple, in this chapter I'll assume that you are interested in assembling a peer-to-peer network (or already have such a network that you'd like to add Wi-Fi capabilities to).

Hubs

A hub is a wired device that is the simplest way to connect three or more devices. A *hub* is basically a box that networked computers connect to via several ports. The hub simply replicates the signals coming into each of its ports and sends the signals to each of its other ports. This is another way of saying that the hub receives information from any device plugged in to it and transmits the information to all other connected devices. It neither knows nor cares which devices the information is going to. It is up to each individual device to pick up the data meant for it. Plugging four devices in to a hub has more or less the same result as connecting the devices to each other.

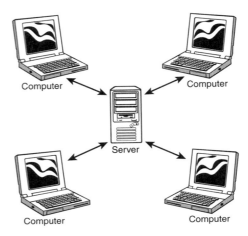

FIGURE 11.3

In a client/server network, a central server controls the resources shared by client computers.

Typical wired hubs are very inexpensive and come with four or five sockets for connections to computers, but some hubs can have a great many more connections.

Switches

You might have heard the term *switch* in connection with networks. A switch is just an intelligent hub. Like a hub, a switch is a device used to connect computers. But a hub has no smarts and simply replicates the signals coming in from each computer and passes the signal along to all the connected computers. In contrast, a switch has built-in "intelligence" that understands where to send transmissions.

Small networks usually don't need switches. The busier the network becomes, the more important it is to use intelligent switches rather than hubs.

These days, even lower-end hubs tend to have some intelligence built-in and are called switches.

Figure 11.4 shows an inexpensive switch in use as a simple hub.

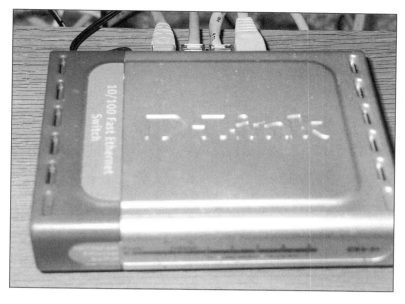

FIGURE 11.4

An inexpensive hub/switch in use.

Routers

A router sits between one network and another. If you are interested in setting up a small home or office network, you are likely to use a router to connect the Internet (the largest network of all) with your small network.

Let's suppose that you have a cable modem, or a DSL modem, connected to the Internet at your home. The router connects to the modem and also to your home network, as you can see in Figure 11.5.

You should know that most routers also function as hubs/switches and provide four or five wired connection sockets. As I'll discuss shortly, routers also come with Wi-Fi. So it's very common these days to buy one inexpensive, little box that combines the features of a wired router with those of a Wi-Fi access point.

If you only plan to connect a few devices to your network, in addition to your modem, you probably only need a router. You can always add hubs as you need them.

Figure 11.6 shows a wired router in use.

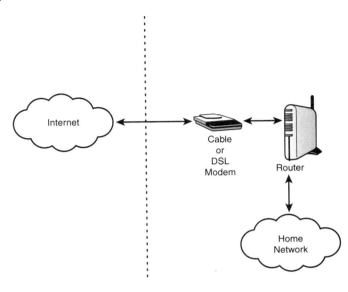

FIGURE 11.5

The router sits in between the Internet connection and your home network.

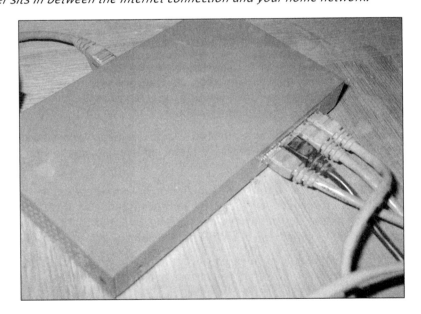

FIGURE 11.6

This wired router distributes an Internet connection across a small network.

As I mentioned, it's common to combine wireless capabilities with those of a router. Figure 11.7 shows a device that combines a wired router with Wi-Fi (as you can tell by the antenna on the upper right, and the networking cables that are plugged in to the device).

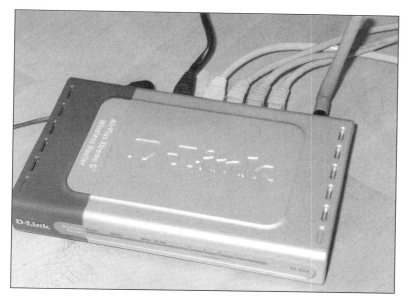

FIGURE 11.7

This device combines a wired router with Wi-Fi wireless broadcast capabilities.

Besides their function as a kind of gateway between networks, most routers provide some additional functionality. Routers can make Internet access possible by translating local network addresses to ones that work on the Internet, a feature called Network Address Translation (NAT), and by assigning network addresses to local machines on-the-fly. This enables the computer (or computers) that make up your home network to interact with servers on the Internet.

Most routers also include features that protect your data by blocking some kinds of information from accessing your network using what is known as a *firewall*. A firewall is a blocking mechanism—either hardware, software, or both—that blocks intruders from accessing a network or individual computer. For more about using personal and network firewalls, see Part V, "Securing Your Computer and Network."

What Is the Network?

How many stars are there in the sky at night, and how vast is the network? Before I wax too poetic, let me get to the point!

Small networks are created by connecting devices, usually via a hub. Larger networks are simply aggregations of small networks, with the small networks connecting to each other, and to the Internet, via routers.

Although these building blocks are very simple, it is obviously possible to create complex network topologies, or arrangements of a network, using them. There are in fact infinite varieties of the possible ways to arrange networks.

You'll have to fit the network topology you create to your physical needs. How many computers do you need, and where?

As an example, Figure 11.8 shows a fairly simple network topology that has seven connected devices and uses a router and two hubs.

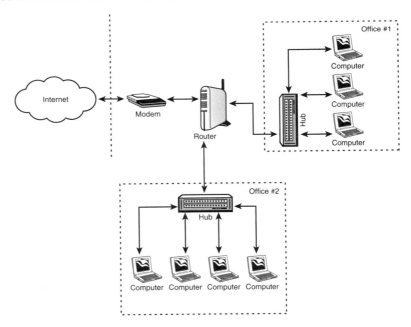

FIGURE 11.8

You need to create a network topology based on your physical requirements.

It's worth stepping back for a second and thinking about what you are creating when you string together computers to make a network. By combining computers in a network, you've created a new entity (the network) that has more computing power than any individual device on the network. Hallelujah! It gets even better when you use a router to connect your network to even more powerful external networks such as the Internet. Effectively, you've harnessed the power of many discrete networks running all over the world in every node of your home or small office.

When this kind of network has become commonplace (or "ubiquitous" as they say in marketing departments), what really is the computer? It doesn't seem right to think of the computer as limited to a single CPU (or box and monitor) when the computer is functioning as part of a network. Maybe the network really is the computer. How much more powerful, and easy, it all becomes when you don't need wires to make the network connections, and you can take your computer (for example, the network) everywhere you go.

Different Ways to Use Wi-Fi in a Network

In the last section, I discussed some of the implications of "stringing together" computers to make something more powerful than any individual computer on the network. But suppose that you didn't have to physically "string" the computers? Well, of course, with Wi-Fi you don't.

If you are going to Wi-Fi–enable a small network, there are really two ways to go about doing it:

- You can use Wi-Fi to connect the entire network.

- You can mix wired and Wi-Fi connections.

Of course, the option of creating an entire network without wires using Wi-Fi is probably only available if you don't already have a network. If you already have a wired network that works, you most likely will not want to replace it. Instead, you'll just want to add Wi-Fi capabilities so that you can go mobile.

For a completely Wi-Fi network, assuming that you want to have a shared connection to the Internet, you still need a cable or DSL modem for that purpose. In addition, you'll need a Wi-Fi router to connect to the modem.

A Wi-Fi network of this sort will look something like the network shown in Figure 11.9.

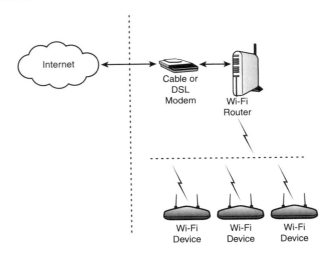

FIGURE 11.9

This network topology shows a setup that uses only Wi-Fi—and no wires—to connect to the Internet.

Perhaps you already have a working wired network. It probably makes sense to keep it. After all, there's a great saying, "If it ain't broken, don't fix it." Besides which, it probably doesn't make sense to retrofit your network devices for Wi-Fi. You simply might not be able to retrofit some devices, such as older network printers, to work with Wi-Fi. You might not want to go out and buy all new equipment that works with Wi-Fi.

> **CAUTION**
> In some cases, household devices such as microwaves and telephones have spectrum conflicts with Wi-Fi networks. This can be another reason to maintain some wired connections in your network.

If you are adding Wi-Fi capability to an existing wire-line network, what you have to do is connect a Wi-Fi access point (AP) "after" the router. ("After" means that you need to add the access point on your side of the router, not the Internet side.)

Figure 11.10 shows what a small network with both wired and Wi-Fi capabilities might look like.

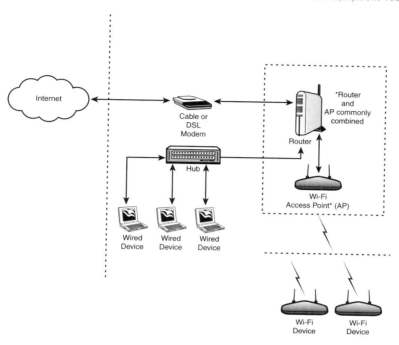

FIGURE 11.10

In mixed wire-line and Wi-Fi networks, the Wi-Fi access point needs to be connected "after" the router.

The Equipment You'll Need

Clearly, the equipment you need depends on how you want to set up your network. If you are planning to add Wi-Fi to a wired network, in addition to all the equipment you need for the wired network, you need a Wi-Fi access point. As I've mentioned, you can also buy a single unit that works as a router for both wired and Wi-Fi computers.

In either case, you'll need Wi-Fi capability on each device that you want to connect to the Wi-Fi network. Of course, you don't have to worry about this with a Centrino laptop—because wireless connectivity is already on board. You should know that it is easy to add Wi-Fi capabilities to other, older computers—both laptops and desktops—simply by adding a Wi-Fi card.

If you are not adding Wi-Fi to your network, but rather planning to create a completely wireless Wi-Fi network from scratch, you need a Wi-Fi router to connect to your modem.

What's It Going to Cost?

> **NOTE**
>
> Although you can buy a DSL or cable modem that is integrated with a Wi-Fi router as a single unit, this is not that common and has a couple of drawbacks—namely a probable higher cost and less ease of upgrading the modem.

It's really dangerous mentioning dollars-and-cents figures in a book for several reasons. The cost of computer-related devices is constantly changing, mostly going down over time as specific technologies become more mainstream. By the time you read this book, it's a safe bet that it will cost less to create a Wi-Fi network than it cost as I was writing the book. (And it was inexpensive enough then.)

It is also true that there are many different possible network configurations, as explained earlier in this chapter. Your network is unlikely to be my network. The cost of setting up your Wi-Fi network will largely depend on how many devices you need to equip with Wi-Fi.

All that said, the short answer is that it won't cost you much. It's truly inexpensive to set up a small wireless network capable of serving a home or small office. (Commercial grade equipment that can handle many multiple users is a different story.)

As of right now, you can buy an excellent wired router and 802.11g Wi-Fi access point for about $50. In addition, you'll need your laptop or laptops that use Intel Centrino mobile technology. If you don't already have an Internet connection, you'll need to pay for a connection, and buy (or lease) a cable or DSL modem. Finally, expect to pay about $30 each for cards used to add Wi-Fi to older desktop and laptop computers.

Summary

Here are the key points to remember from this chapter:

- Networks sound complicated, but they are really simple to understand when you realize that they are made up of small building blocks that are all more or less the same.

- You can create entirely Wi-Fi wireless networks or add Wi-Fi wireless capabilities to an existing network.

- In most cases, you'll want your Wi-Fi network nodes to connect to ordinary DSL or cable modem Internet access.

- It's inexpensive to put together a Wi-Fi network or to add Wi-Fi capabilities to an existing network.

Buying a Wi-Fi Access Point or Router

Perhaps you've used your laptop with Intel Centrino mobile technology on the road and have seen how great wireless and Wi-Fi are. It's also okay if you haven't. (Maybe you don't like to travel.) In any case, you are ready to "unwire" your home or small office by adding a wireless Wi-Fi network you can use with your Centrino laptop and other devices that have been enabled for Wi-Fi.

You'll find that it's very easy to create a Wi-Fi network from scratch (or to add Wi-Fi to an existing Ethernet network).

This chapter gets you started building your own Wi-Fi network. There are quite a number of different pieces of hardware (or hardware combinations) you can use to do this, which can seem a bit overwhelming. But, fear not! It's all a lot simpler than it might seem.

I'll start out by clearly delineating the different kinds of hardware that you might use in creating a home or small office Wi-Fi network. There's no need for any confusion about this. Then, I'll take a look at all-in-one Wi-Fi networking kits.

Finally, I'll show you some of the most popular access points: the Apple Extreme and AirPort Express base stations and combo wire line—Wi-Fi routers from D-Link and Linksys.

Understanding the Different Pieces of Hardware

So you want to set up a new wireless network in your home or office. Or, you want to extend an existing wire line network to provide wireless capabilities. In either case, relax! These are pretty easy things to do using Wi-Fi technology. You'll have your wireless network up and running in no time—and you will be using your computer in your living room, in your garden, or on your deck, as well as from all kinds of unlikely places.

This chapter focuses on buying the right hardware to set up your wireless network. (Subsequent chapters detail other aspects of setting up and administering your wireless network.)

Networking hardware can be somewhat confusing and uses terms that overlap. Before I get down to specifics at a brand-name level, it's a good idea to spell out the different kinds of equipment that you might expect to need in a wireless network. If you take one thing away from this book, let it be clarity about the different pieces of wireless networking hardware and what they do.

WHICH STANDARD SHOULD YOU BUY?

This chapter does not discuss in detail the differences between the various Wi-Fi protocols (principally 802.11b and 802.11g) because they are amply discussed in Chapter 8, "Entering a World Without Wires," and Appendix A, "Wireless Standards." But you should bear the different flavors of Wi-Fi in mind when you decide which equipment to buy. 802.11g equipment is now the hands-down choice for the following reasons:

- 802.11g is almost five times faster then the older 802.11b standard

- The price of 802.11g equipment is about the same, and in some cases 802.11g products are now less expensive than 802.11b equivalents

- 802.11g products can have better range than 802.11b

- 802.11g products often have better interoperability with equipment from other vendors

- Many 802.11g products from major manufacturers are officially "Centrino Verified" for compatibility (look for a Centrino sticker on the box)

- And last but not least, 802.11g is backward compatible with 802.11b, so if you have some older 802.11b equipment already you can still use it with the new stuff

Ad Hoc Versus Infrastructure Modes

A basic distinction within Wi-Fi networks is between *ad hoc* and *infrastructure* modes. In ad hoc mode, there is no central server, and wireless nodes (meaning computers or other devices) communicate directly with each other on a peer-to-peer basis. Figure 12.1 shows an example of ad hoc wireless networking.

In contrast, in infrastructure mode, the wireless network consists of at least one access point (sometimes called an *AP* for short) connected to a wired network as shown in Figure 12.2. The access point serves as a connection between the wired network and the wireless network, and provides a mechanism for wireless computers on the network to share access to resources such as files, printers, and Internet access. As you'll see later in this chapter, with the Apple AirPort Express, an access point can even become a way to distribute music files in your home.

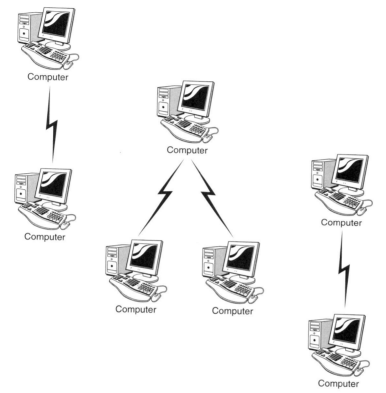

FIGURE 12.1

These computers are networked using wireless ad hoc mode.

The main problem with ad hoc mode is that as implemented it doesn't have very much range. Access points are simply better at broadcasting Wi-Fi signals than Wi-Fi cards or chips within a computer.

If you only have two computers that are not physically far from each other, you might want to try ad hoc mode and skip the network hardware explained in this chapter. (Ad hoc mode can also be useful when you are on the road and want to connect to a colleague's computer.)

Ad hoc mode can also be used to make a desktop PC wired to the Internet into an access point for another computer, assuming that both have the appropriate Wi-Fi cards. This requires that the computer wired to the internet be configured to share its Internet connection, as I'll explain in Chapter 14, "Configuring Your Wi-Fi Network."

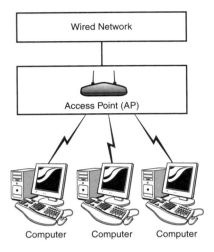

FIGURE 12.2

In infrastructure mode, at least one access point connects the wireless and wire line portions of the network.

Using computers like this in the ad hoc mode eliminates the need to purchase an access point or router (not that much of an expense these days really), but is more complicated to set up and doesn't eliminate as many wires. In addition, ad-hoc mode may be less secure.

In Chapter 14, I'll show you how to configure a wireless computer to enable ad hoc networking and also how to share an Internet connection. It's interesting that extensive use of ad hoc networking, combined with the use of access points, can create a powerful network of wireless computers configured in what is known as a *grid*—provided that enough computers are involved.

> **CAUTION**
>
> Note that unless one of the computers involved in an ad hoc connection has a separate non–Wi-Fi connection to the Internet, neither computer will be capable of accessing the Internet.

For the most part, you are less likely to be frustrated if you go ahead and set up an infrastructure mode wireless network, which is my assumption for the remainder of this chapter.

Wi-Fi Access Points

The crucial piece of equipment in your Wi-Fi network is the access point, also sometimes called a *base station*. A Wi-Fi access point

- Transmits (and receives) Wi-Fi broadcasts

- Acts as a connection between a wired computer and the Wi-Fi network (this functionality is call "routing," so the the part of the base station that performs it is usually called a "router")

Because Wi-Fi access points come in a number of different permutations, it makes sense to take a bit of a closer look.

If you already have a wired network, you are probably using a router, which is a small device that connects to your broadband Internet modem and is connected to each of the computers on your network as shown in Figure 12.3. (Don't worry, we'll get to wireless networking in a moment. The point here is to be sure that you understand how wired networking works before you "throw away" the wires.)

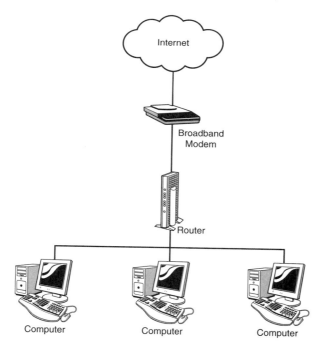

FIGURE 12.3

A router sits "between" your Internet connection and the computers in your network.

> **NOTE**
> Although your network router is most likely a small, separate device, you might be interested to know that server computers can be used for the same purpose.

The router provides the services necessary for each computer on the network to have Internet access. It also can assign each computer on the network an Internet Protocol (IP) address so that the computers can communicate and share resources.

Small routers typically come with four or five sockets for computers to plug in to. If you need to connect more computers, you can easily add hubs, which are simply wired network repeaters.

Most Wi-Fi access points provide the same services as a router, so you really don't need a router any more. (But you'll need to make sure that your Wi-Fi access point can function as a router.) You can eliminate your router and instead plug in the Wi-Fi access point (which is really a combination access point and router).

The best part of it is that many access point units, such as the D-Link and Linksys units described later in this chapter, provide both Wi-Fi access for network computers and plugs for the wired devices on your network. A network arranged using an access point/router that provides both Wi-Fi and wired connections is shown in Figure 12.4.

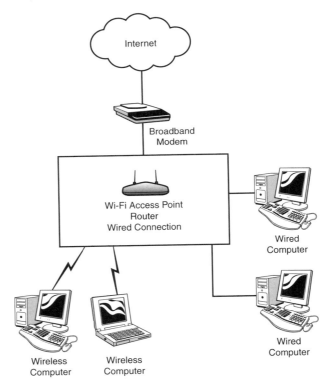

FIGURE 12.4

Some access points provide Wi-Fi and wired network connections.

Simplicity is a good thing, and—if you are like me—it's great to get rid of all those small devices cluttering up your life near the Internet connection. (In my case, that's under my desk!)

You can take the combination thing one step further and buy a small unit that combines broadband modem with a Wi-Fi access point and router.

For example, if you are using a cable modem to access the Internet, you could replace your cable modem with the all-in-one SURFBoard Wireless Cable Modem Gateway SBG900 from Motorola, which includes a modem, router, wireless access point, and five sockets for wired network connections. This unit runs 802.11g Wi-Fi and can be had for a bit less than $200 U.S. (Other wireless access point cable modem combos can be had for closer to $100, such as models from Netgear.)

> **TIP**
>
> If you want to keep your wireline router, you can. Depending on your network configuration, it might make more sense to plug your access point in "after" the router, rather than removing the router. As I'll discuss in Chapter 15, there are many possible ways to set up a network.

If you make this substitution, the physical network shown in Figure 12.4 can be redrawn more simply, as shown in Figure 12.5.

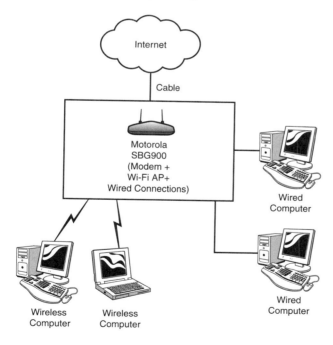

FIGURE 12.5

It's possible to buy a combo cable modem unit such as the Motorola that includes a router, plugs for network wires, and a Wi-Fi access point.

Wireless Bridges

You should also know about another kind of device that can be used to extend your Wi-Fi networks, or to make them more versatile. This is the wireless network bridge.

Wireless network bridges come in three flavors:

- Simple Ethernet to wireless
- A dedicated wireless bridge
- Wireless bridge access point combinations

A simple Ethernet-to-wireless bridge is intended to connect a wired device to a wireless network, and you can use it to connect any device capable of networking—such as a network printer or a game box with an Ethernet port—to a wireless network.

A decent Ethernet-to-wireless bridge such as D-Link's DWL-G810 can be had for a little more than $90.

Dedicated wireless bridges can be used to connect a wireless network to a wired network. They can also be used to extend the range of a wireless access point. When used this way, they are sometimes called *repeaters*.

High-end dedicated wireless bridges, sometimes also called *workgroup bridges*, include management and security features and can be fairly expensive. But for home or home office use as a repeater, you should be able to get a dedicated wireless bridge for about $120. A good choice at about this price point would be Linksys's WET54G Wireless-G Ethernet bridge, which can be used to bridge or to repeat 802.11b and 802.11g Wi-Fi networks.

USING YOUR MOBILE COMPUTER AS A WIRELESS BRIDGE

You can also configure your laptop that uses Intel Centrino mobile technology to work as a wireless bridge, although, of course, it will lack the power and range of either a dedicated hardware bridge or an access point.

Obviously, the mobile computer must be in place—meaning, not very mobile—to work as a bridge or a repeater.

For the details of how to simply add the wireless capabilities of your Centrino laptop to a wireless bridge using Windows XP, see Chapter 3, "Configuring Your Mobile Computer."

In addition, some access points can also be configured as wireless bridges. When an access point is configured this way, it cannot also be used as a normal access point. (An exception to this statement is that some high-end enterprise class access points can be simultaneously used as bridges.)

An access point configured as a wireless bridge is probably the least expensive repeater you can buy. If you want one of these, you can have one for as little as about $70. A decent example is Linksys's WAP54G.

> **TIP**
>
> If you are looking to buy an access point that can be configured as a wireless bridge to use as a repeater, check that the product specifications say something similar to "Wireless Access Point Roaming and Bridging."

Wireless Networking Kits

A number of companies produce wireless networking kits. These kits are essentially a bundle containing an access point, a wireless PC Card or wireless USB connector, software drivers, and instructions.

These network kits cost less than $100 at a discount retailer. The Netgear WGB511 is an example. It bundles an 802.11g access point/router with an 802.11g PC Card.

Nothing is particularly wrong with these networking kits, but they don't get you very far—particularly if you already have a laptop using Intel Centrino mobile technology. Because you already have wireless connectivity for this mobile computer, you certainly don't need a wireless card or USB connector for it.

Don't get me wrong—I'm all for anything that will make life simpler for you. But in this case, wireless networking kits don't bring much to the party. You can start with one of them and expand your network later if you want. Still, it is no more complicated to buy an access point on your own and create your own "kit."

There is one situation in which a wireless networking kit does make sense. Let's say you have an old laptop in the house in addition to the new Centrino-based computer you just purchased. You would like to give the older laptop to a spouse or your kids, and of course want it to be connected to the Internet by wireless. The bundle gives you the access point/router you need in any case, and the card for the old laptop.

Choosing a Wi-Fi Access Point

A good 802.11g Wi-Fi access point can be had for between $50 and $100. If you are willing to settle for 802.11b, you can get one for well under $50. By contrast, an elegant 802.11g unit, the Apple Extreme Base Station, costs about $250 ($200 if you take the model without the external antenna port). Although industrial-strength commercial units can cost a good bit more, the point is that these are not hugely expensive pieces of equipment.

You'll pay more for 802.11g equipment than for 802.11b equipment because 802.11g is newer and faster. This is a choice with obvious trade-offs that you'll have to make, but at this point the balance has pretty much tilted toward 802.11g because the pricing has come down so much.

Likewise, the Apple Extreme Base Station costs a little more than equipment manufactured by a vendor that is not Apple—but then again, it looks so cool. You'll have to decide how much the cool factor matters to you.

> **NOTE**
> Be sure to buy the equipment you need. You can buy a wireless access point that also includes the functionality of a router (for example, the D-Link DI-624). Some wireless access points *do not* provide the router functionality, so if you already have a router you can buy one of the wireless access points that does not include a router (for example, the Linksys WAP 54G).

Even if you buy your access point in the real world of bricks and mortar, it makes sense to at least comparison shop online.

I'm a firm believer in buying equipment from quality vendors that stand behind their products. Quality manufacturers of Wi-Fi access points include

- Apple
- Belkin
- D-Link
- Linksys (owned by Cisco)
- Netgear
- SMC Networks

Specific Brands and Models

I'd like to show you a little more about units from Linksys, Apple, and D-Link. In Chapter 13, "Setting Up Your Access Point," I'll show you how to configure these units in your network.

> **NOTE**
> Most Linksys wireless broadband routers have been officially tested and verified for use with Intel Centrino mobile technology. The latest Linksys wireless routers support Intel Smart Wireless Solution technology, which simplifies setup and security.

Your laptop computer doesn't care whether the access point has an Apple nameplate. It will connect to the access point using Wi-Fi just as easily. In other words, Wi-Fi is operating system neutral. You can choose to run Windows XP, Linux, or Apple OS-X, and that is, as my Grandma used to say, "your nevermind."

Linksys Wireless Broadband Router

Linksys's Wireless Broadband Router (model BEFW11S4 runs 802.11b; model WRT54G runs 802.11g), shown in Figure 12.6, is an access point which you should be able to buy for a street price of around $40–80.

Note the two antennas on either side of the unit shown in Figure 12.6. Having two antennas like this helps give this unit solid signal strength and range.

The Linksys unit provides a connection to a modem connected to the Internet as well as four wired Ethernet sockets. These connections are shown in Figure 12.7.

FIGURE 12.6

The Linksys Wireless Broadband Router is decent and inexpensive.

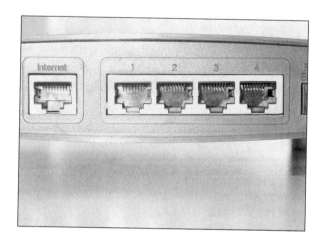

FIGURE 12.7

The Linksys Wireless Broadband unit provides four wired network sockets.

As you can see in Figure 12.7, you can plug the Linksys Broadband Router in to your cable or DSL modem, connect your wired computers to it, and use the unit to connect via Wi-Fi to wireless computers without needing further hardware. With this unit, you don't need no stinking router, or even a switch or hub (provided you have four or fewer wire-line computers).

The Apple AirPort Extreme Base Station

When I look at the AirPort Extreme Base Station, shown in Figure 12.8, I see something that somehow manages to combine cute with high tech. I don't know whether you've ever seen Woody Allen's movie *Sleeper*, but it reminds me of something from that retro *Back to the Future* film.

FIGURE 12.8

The Apple Extreme Base Station looks like a prop from the movie Sleeper.

The Apple Extreme Base Station might look cute, but it also has quite a bit of brains and brawn behind that exterior. For one thing, it runs 802.11g, so it is fast. You can tell simply by looking at the sockets on the back of the unit that it delivers some nifty features. If you look at Figure 12.9, you'll see the sockets on the back of the Apple Extreme Base Station.

A couple of the sockets shown in Figure 12.9 bear some elaboration. The external antenna port and the connection for a dial-up modem are only available on the more expensive of the two AirPort Extreme models available. The Ethernet WAN (or wide area network) port is used to connect to a cable or DSL modem (or, in some cases, to a network connected to the Internet). The Ethernet port can be used to connect a wired network "behind" the AirPort Extreme Base Station. You can connect the port to a hub or switch, and then add quite a few wired devices. That way, you wouldn't need a separate router. The USB printer port provides a way to give all the wireless devices connected via the AirPort Extreme Base Station access to a printer with a USB port.

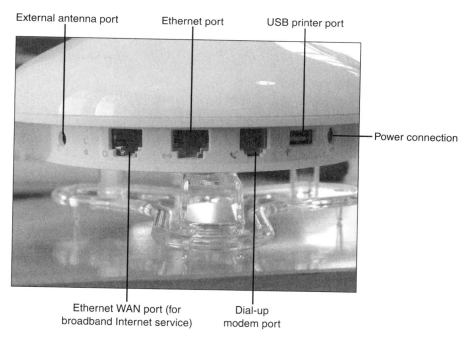

External antenna port   Ethernet port   USB printer port

Power connection

Ethernet WAN port (for broadband Internet service)   Dial-up modem port

FIGURE 12.9

By looking at the sockets on the back of the AirPort Extreme Base Station, you can see that it provides extra functionality.

The AirPort Express Base Station

The AirPort Express is a small 802.11g access point with some unusual features. Figure 12.10 shows the AirPort Express and its connectors.

The feature of the AirPort Express that has received the most notice is that it can easily serve iTunes musical files through any stereo in your house—or simply through a set of speakers. You do have to connect the stereo to the AirPort Express (through the audio mini jack). This might require additional cables; you can

> **TIP**
> The AirPort Extreme Base Station provides a standard antenna connector, so you can add any antenna you'd like (and pay less than at an Apple company store). For more information about antennas, see Chapter 16, "Adding Wi-Fi Antennas to Your Network."

buy a set from Apple for about $30. It also has a built-in USB port, so you can use wireless networking to share printers.

The AirPort Express can be used as a handy and convenient way to extend the range of a Wi-Fi network created using a larger access point (such as the AirPort Extreme Base Station).

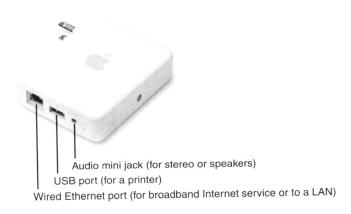

Audio mini jack (for stereo or speakers)
USB port (for a printer)
Wired Ethernet port (for broadband Internet service or to a LAN)

FIGURE 12.10

The AirPort Express Base Station combines a small size with some unusual features.

TIP
You can also buy dedicated wireless media streamers, which use Wi-Fi to route music, video, and photos stored on a computer to your stereo or television; for example, the SMC EZ-Stream Wireless Multimedia Receiver runs over 802.11a, 802.11b, and 802.11g and sells for about $200.

It's convenient to travel with an AirPort Express unit. That way, if your hotel room is only equipped with an Ethernet Internet connection (and doesn't provide wireless), you can plug the AirPort Express in and use your mobile laptop from anywhere in the room without having to worry about wiring.

The only disadvantage of the unit is the single 10/100BASE-T wired Ethernet socket. This can be used to connect to the Internet or to a router or hub that is part of a wired network. But the AirPort Express can only do one or the other; so if you want to integrate it as part of a wired network, you will need additional hardware such as a router.

D-Link XTreme G Wireless Router

The D-Link XTreme G Wireless Router model, DI-624, is probably the most popular combination wire line router and 802.11g Wi-Fi access point. This model combines speed and features with a low price (about $50 from a discount retailer including rebates).

Figure 12.11 shows the status panel of the D-Link XTreme.

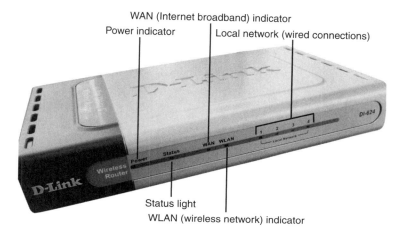

WAN (Internet broadband) indicator

Power indicator

Local network (wired connections)

Status light

WLAN (wireless network) indicator

FIGURE 12.11

The indicator lights tell you which of the D-Link XTreme features are currently in use and working.

It's worth having a look at Figure 12.11, which I've marked with the purpose of each of the indicator lights shown. Haven't you ever been curious about what all the lights on one of these units actually means? If you are trying to debug a Wi-Fi access point or a router, it's very helpful to know what they show you.

The power indicator means that the unit is connected to electrical current. If all is working properly, the status light should be blinking. The WAN (Internet connection) light blinks to indicate that it is working. The WLAN (wireless network) light blinks when Wi-Fi radio transmissions are being sent or received. Each of the local network lights are on solidly to mean that a wired connection has been made in the corresponding plug.

NOTE You can buy dedicated wireless servers specially intended to share printers with a wireless network. These servers provide a USB port to connect to the printer. Models such as the Belkin Wireless Print Server sell for around $100.

Figure 12.12 shows the connection sockets for the D-Link Xtreme unit.

As you can see in Figure 12.12, the D-Link XTreme can be connected to a broadband modem, can broadcast wireless signals, and, at the same time, it can provide the ability to plug up to four wired computers using standard Ethernet. Of course, there's nothing to stop you from plugging hubs or switches in to the unit, easily expanding your network to more than four wired devices.

Local network (wired Ethernet connection) plug #4

Local network (wired Ethernet connection) plug #2

WAN (Internet broadband) connection (for cable or DSL modem)

Power adapter

Local network (wired Ethernet connection) plug #1

Local network (wired Ethernet connection) plug #3

FIGURE 12.12

The D-Link XTreme provides connections for wired Ethernet and to a broadband modem (as well as Wi-Fi capabilities).

Summary

Here are the key points to remember from this chapter:

- You should understand what the different pieces of wireless networking gear do before you buy one.

- An access point, also called a base station, is the heart of a wireless network.

- A wireless bridge can be used to connect a device with an Ethernet connection to a wireless network.

- A wireless bridge can be used to extend the range of wireless access points.

- Home and small office wireless access points are much less expensive than they used to be and range in cost from less than $50 to around $250.

Setting Up Your Access Point

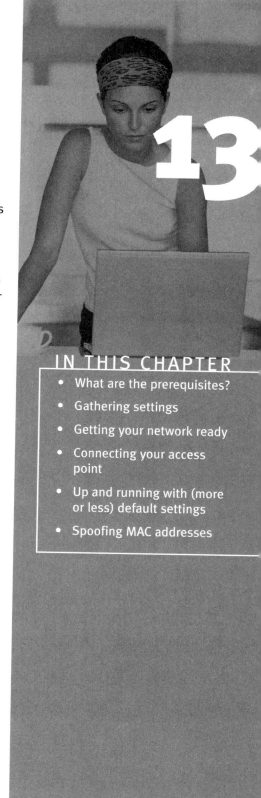

S o you've taken the plunge and bought a Wi-Fi wireless access point! Congratulations! This chapter tells you how to plug it in and use it both to create a network and to connect to the Internet.

What are you waiting for? Creating a wireless Wi-Fi network is usually quick and painless. This chapter provides the information you need to make it a painless experience.

Preliminaries

It's good news that usually a standard installation of a wireless access point is not much harder than plugging it in and turning it on. Typically, you'll be ready to use your wireless network (and shared Internet connection) within minutes.

But, much as I love Wi-Fi wireless technology, I have to be honest. There can be a dark side to setting up a Wi-Fi access point. It's as if you were walking along a mountain path with steep drops on either side. As long as you can keep to the path, everything is fine; but if you slip off either side, you can run in to trouble.

Trouble tends to come in two forms:

- If your network is at all complicated, configuring your access point might require a little moxie.

- Access points are tricky pieces of equipment, and a small percentage of them simply ship from the factory with defects.

Regarding the first kind of trouble, network complexity, you probably won't get into it. That is, unless you already have a complicated network, or need to set up a complex network. In this case, you'll find more information in Chapter 14, "Configuring Your Wi-Fi Network," and in Chapter 15, "Advanced Access Point Configuration." Chapters 14 and 15 provide a great deal of information about different ways to set up networks, as well as some of the choices you can make regarding how to design and set up a network.

Perhaps there isn't a whole lot to be said about defective equipment. Every category of electronic device can have an occasional lemon. Your take-away from this should be that if you are following the manufacturer's directions—and the advice in this chapter—and your access point is just not working, there is a possibility that the problem is with the equipment.

Here are the general steps you need to do to get your access point working to form a network with your computers and to provide shared Internet access for the computers on the network:

> **NOTE**
> If you are using encryption (as you should) and you have connection problems from your Centrino laptop, the problem is most likely not defective equipment but rather with the encryption key and settings (also called the "security setting"). This can be frustrating to deal with. The thing to do is to go back and check the value of your key and related settings as entered in the access point and carefully replicate them using your laptop. It's also a good idea to check the support section of the equipment vendor's website for known problems and information about how to fix them.

- Collect information about the settings necessary to connect to the Internet (see "Collect Your Settings" later in this chapter for information about which settings you need to note).

- Position your access point. Wireless access points can be very sensitive to where they are located and to interference (see "Plugging in the Hardware" later in this chapter for some suggestions about how to locate your hardware).

- Connect the access point to your cable or DSL modem.

- Plug in the access point.

- Configure the access point (see "Configuring Your Access Point" later in this chapter for the details).

- Set your wireless Wi-Fi devices to communicate using the access point.

- If you have any wired devices using the access point/router, you'll need to make sure that their settings reflect the new hardware.

Voilá! It's not very hard, and you'll be up and running with your wireless network very quickly—usually in less than half an hour.

> **NOTE**
> Some access points are not set up quite in this order. If the directions from the manufacturer differ from the steps I've outlined, you should, of course, follow their directions.

For the most part, in the remainder of this chapter, I'll assume that you are creating a network from scratch using your wireless access point (which I'll also assume includes router functionality). This is probably the most common situation for a Wi-Fi beginner and, in many ways, the most straightforward setup scenario.

I'm also assuming the straightforward and simple Internet connection settings. In other words, I can't go into details for every obscure way to connect to the Internet. (For example, Australia's Big Pond cable has some special access settings.) The thing to bear in mind is that connecting a wireless access point and router to the Internet uses exactly the same settings for the Internet connection that your broadband Internet provider (ISP) gave you to connect a single computer. So if you have any problems with the Internet settings for the access point, you should verify them with your ISP by contacting their customer support department.

ADDING WI-FI TO AN EXISTING WIRED NETWORK

If you are adding Wi-Fi to an existing wired network with a router, from a hundred-mile view, there are two ways you might go about it:

- Replace the router with the router functionality built in to the Wi-Fi access point

- Add the access point "behind" your existing router

If you are replacing an existing router, you'll need to plug your current wires in to the Ethernet socket (or sockets) provided on your new access points/router (or replace your wired network cards (NICs) with Wi-Fi cards so that you can do away with wires altogether).

You Need a Working Internet Connection

It might seem obvious, but the first prerequisite to setting up a Wi-Fi access point—that will let the computers on your network connect to the Internet—is a working Internet connection. This is true for the obvious reason: You can't connect to the Internet without a connection. But a couple of other points are involved as well. You'll need to know some of the settings to put into the access point so that it can connect to your Internet service provider (ISP).

One of the easiest ways to find these settings is to read them off your Internet-connected computer. (See the following section, "Collect Your Settings," for details.)

Many cable and DSL service providers might state as their official policy that they only support one computer connected to their service. But service providers have come to realize that a great many people connect networks, including wireless networks, to the Internet. If you have a network attached to the service, they won't help you configure the network—at least not for free—but they will help you with your Internet connection. To do this, quite likely they will ask you to connect a single computer to the modem attached to your broadband Internet connection.

All this means is that, if at all possible, you should have a computer up and connected by wire to the Internet first to collect the Internet settings, and you should also make sure that the connection is working properly.

Collect Your Settings

You need to know the settings you should use to tell the access point how to connect to the Internet. The easiest way to find these settings out is to read them off your computer or to learn them from your ISP.

Collecting setting information is a little different in Windows XP as opposed to Windows Me, so I'll show you each. (This works the same way in Windows 2000 as in Windows XP, and Windows 98 is no different from Windows Me in this respect.)

> **NOTE**
>
> This book mostly assumes that you are running one of the versions of Windows XP because that is what will be installed on a laptop that uses Intel Centrino mobile technology. The other operating systems are discussed in this chapter because your older, wired computers connected to the Internet might be running them—so you might need to use them to collect information.

Please be very clear that the point of opening up these dialogs is to see how they are set so you can configure your access point/router to connect to the Internet. It might help to think of your network as having an inward and an outward face. Right now, we are gathering the settings for your network's outward face—the part connecting to the Internet, sometimes called the WAN (or wide area network).

The inward face of your network consists of your connected computers, or LAN (local area network). When the access point/router is in place, and it's time to verify settings of the computers within your LAN, all the computers that share access through it should be set to automatically obtain IP and DNS server addresses. (These are obtained from the access point/router.)

Windows XP

From the Windows Start menu, open the Control Panel. In the Control Panel, double-click Network Connections to open the Network Connections, as shown in Figure 13.1.

Highlight your computer's connection to the broadband modem. Most likely, this will be something such as Local Area Connection. Right-click the connection and select Properties from the context menu. The Properties window for the connection will open, as shown in Figure 13.2.

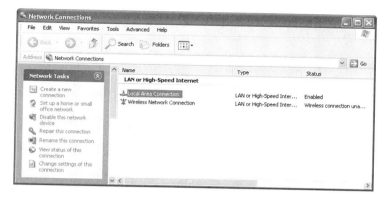

FIGURE 13.1

The Windows XP Network Connections window lets you select a network connection.

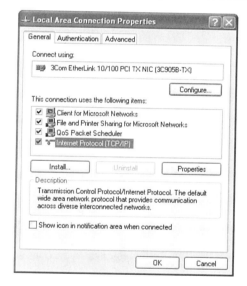

FIGURE 13.2

The Properties window for the Local Area Connection shows a number of items.

In the items list, choose Internet Protocol (TCP/IP) and click the Properties button. The Internet Protocol (TCP/IP) Properties window, shown in Figure 13.3, will open.

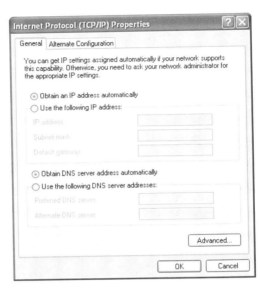

FIGURE 13.3

You can determine your settings for connecting to the Internet using the Internet Protocol (TCP/IP) Properties window for the device that connects to your broadband modem.

The Internet Protocol (TCP/IP) Properties window is where you should note the settings required for your Internet connection. Quite likely, the settings in this window will be set to automatically obtain IP and DNS server addresses as in Figure 13.3, and you should make note if this is the case.

However, it is also common to have to supply specific server addresses to connect to the Internet. Even when the IP address (which represents your particular node on the Internet) is automatically generated, you will often need to specify an address for your DNS server—which translates domain names into the numeric IP address necessary to connect to specific websites.

Figure 13.4 shows a computer connected to the Internet using an automatically generated IP address, but with a specified DNS server address.

Whatever the settings, you should write them down, taking great care to be accurate about the numbers that compose the address of a static IP or of a DNS server. You will need this information to configure your wireless access point/router.

TIP

If you have a static IP, you should also make note of the Subnet mask and Default gateway being used, which you will also see in the Internet Protocol (TCP/IP) Properties window shown in Figures 13.3 and 13.4. You will need these settings as well.

FIGURE 13.4

It's common to have to specify an address for DNS service.

Windows Me

Determining your network settings in older Windows versions works pretty much in the same way as under Windows XP, but the interface is a little different.

Open the Network window, shown in Figure 13.5, either by selecting the Network application in the Control Panel (click Start, Settings, Control Panel), or by right-clicking on the My Network Places icon on your desktop and selecting Properties from the context menu.

In the Network window (make sure that the Configuration tab is selected), scroll down the list of installed network components until you find an entry in the list that combines the TCP/IP protocol and the device (such as a network card) that you use to connect to the Internet. A typical entry of this sort, shown selected in Figure 13.5, combines TCP/IP with a 3Com Etherlink Network Interface Card (NIC).

With the combined protocol and device selected, click the Properties button. The TCP/IP Properties window will open, with the IP Address tab displayed, as you can see in Figure 13.6.

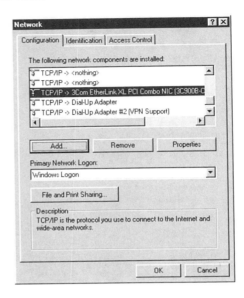

FIGURE 13.5

The Network window is used to select a combination of protocols and devices.

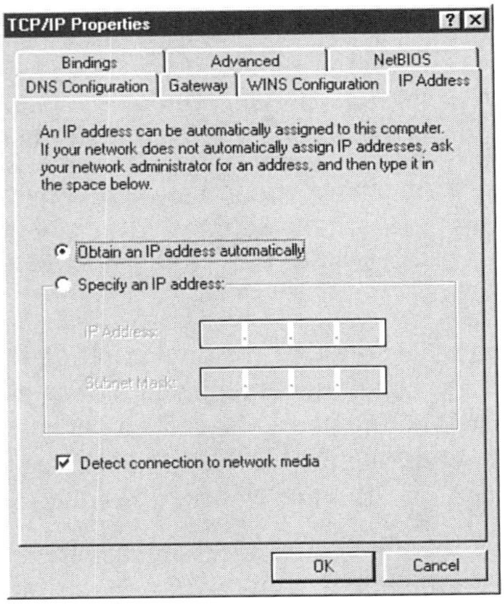

FIGURE 13.6

The IP Address tab of the TCP/IP Properties window shows you the IP settings you should use.

Make a note of the IP settings, and if they are anything other than Obtain an IP Address Automatically, write down the numbers you see.

Next, click the DNS Configuration tab, which you can see in Figure 13.7.

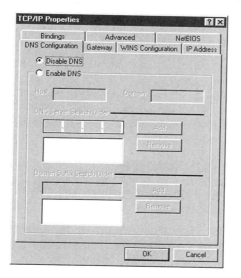

FIGURE 13.7

Make note of the DNS settings you can find on the DNS Configuration tab.

If these settings are anything other than Disable DNS, you should make note of the DNS settings because you will need them to configure the access point/router.

If you are not obtaining IP and DNS settings automatically, you will also need to note the gateway being used, which you can find on the Gateway tab.

SPOOFING MAC ADDRESSES

It's possible that your ISP has registered the MAC—Media Access Control—address of your computer's network card. Each hardware device has a unique MAC address, which is a way to identify the device. Your ISP might control access to the Internet by checking your MAC—and the MAC of your wireless access point router is obviously not the same as that of a network card in a computer connected to your ISP. (They are two different hardware devices.)

If your ISP does this (it amounts to an attempt to regulate the number of computers you have within your LAN connecting to the Internet), you can get around it by "telling" your access point to *spoof* the MAC of your network card—in other words, to adapt the MAC of the registered network card as protective camouflage. For information on how to do this, see "Cloning MAC Addresses" later in this chapter.

As an alternative to spoofing the MAC address, you should consider the straightforward approach: phoning your ISP and requesting that your registered MAC address be changed to that of your router. If your ISP won't do this, you should probably "hire" a different ISP.

Plugging in the Hardware

Now that you've collected your current settings, you are ready to rock and roll...er...plug and unplug things.

> **TIP**
> The port, or socket, for the Ethernet cable on the modem will probably be marked, logically enough, "Ethernet." Ignore the USB socket, which you will also find on many broadband modems, unless you want to plug a USB printer, or other USB device you want to share, in to it.

> **TIP**
> If you're not sure which socket to plug in to on your access point, take a look at the pictures of wireless access point connectors in Chapter 12.

The first thing to do is to power down the computers on your network, as well as any hubs, switches, or routers, and your cable or DSL modem. You might have to unplug the modem to turn it off because many of these devices don't come with power switches.

If an Ethernet cable is plugged in to the back of the modem, unplug it.

Next, using an Ethernet cable (you'll probably find one provided by the access point manufacturer), connect the cable to the appropriate socket on the access point. This socket will probably be labeled "Internet" or "WAN." (WAN is short for wide area network; one way of looking at the Internet is that it is a great, big WAN.)

Now, connect the wired computers and other devices (such as printers) in your network (if you plan to include wired devices) to the appropriate ports on the back of your access point/router, usually numbered 1–4.

If your access point/router has a number of wired sockets, the Ethernet ports might simply be numbered. Otherwise, if your access point has routing capability (as most do today), there will probably be at least one Ethernet socket for your wired network.

If it is a single Ethernet socket, it might be labeled "Ethernet" or "LAN" (short for local area network). On the Apple Extreme Base Station, the Ethernet out port is designated with a special symbol that looks like two angled brackets with dots in between: < ... >.

If there is only one port for your wired devices, or if the number of ports provided on a combo wired-wireless unit is not sufficient, you might have to use a hub, or switch, to add additional Ethernet ports to the wired portion of your network. Don't worry; you can easily find very cheap hubs (or switches) that are fine for this purpose.

> **NOTE**
>
> Try to position the access point somewhere where it is off the floor. If possible, if you have a house with more than one story, the upper floor is good.
>
> The access point should also not be covered by any bulky, particularly metal, objects. Finally, it should be located away from anything that might interfere with its broadcasts, such as microwaves, cordless phones, and other Wi-Fi access points.
>
> Wi-Fi broadcasting can be very sensitive to physical positioning and the materials it is trying to go through. For example, signals fall off radically if you try to go through large house plants (because of their high water content). Because you can't "see" Wi-Fi radio signals, it can be hard to know what is wrong.
>
> If you are having trouble with the range and strength of your Wi-Fi signals, as a first step, you should experiment with repositioning your access point.

When everything is wired up, it should look like the diagram shown in Figure 13.8. The modem is connected to the Internet; the access point is connected to the modem; and the internal, wired network (if there is one) is connected to the access point.

Configuring Your Access Point

When you have your Wi-Fi access point connected to your cable or DSL modem, plugged in, and turned on (as explained in "Plugging in the Hardware"), the next steps are to configure your access point and the computers in your network (sometimes not necessarily in that order).

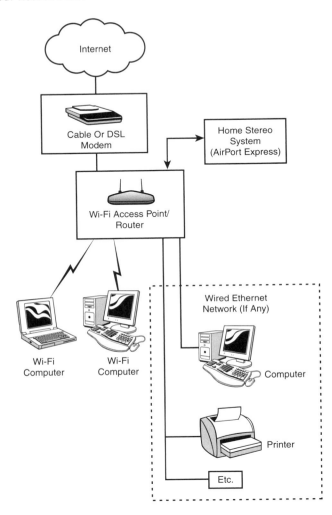

FIGURE 13.8

The Internet is connected to the modem, which is connected to the access point, which is connected to your wired network (if you have one).

Of course, you should follow the directions provided by the manufacturer of your access point device. In this chapter, I'll show you some good illustrative examples of how to get up and running with a Wi-Fi access point. I'll use as demonstration cases Linksys Wireless-Broadband Router, a D-Link AirPlus Xtreme G, and an Airport Extreme Base Station (all described in Chapter 12).

Linksys Wireless-Broadband Router

With your Internet settings collected as explained in "Collect Your Settings," your devices connected as explained in "Plugging in the Hardware," and your Linksys Wireless-Broadband Router turned on, you are ready to configure it.

With a computer connected to the Ethernet port of the Linksys Wireless-Broadband unit, open a Web browser. In the Web browser, enter the address `http://192.168.1.1/`. The username and password box shown in Figure 13.9 will open.

> **NOTE**
>
> Most modern access points and routers are administered with a browser-based interface. The network address of the router varies a bit by router manufacturer. Addresses such as 192.168.0.1 or 192.168.1.1 (as with the Linksys unit) are typical. Check your documentation to be sure in the case of your particular unit.
>
> If you have an older Wi-Fi access point without the capabilities of adding a wire-line computer directly to the unit, you'll need to administer the unit via its wireless connection (follow the manufacturer's directions).

FIGURE 13.9

When you enter the URL for the access point/router in a browser, you will be prompted for the administrator's username and password.

Leave the username blank and enter **admin** for the password, and click OK. The primary administration screen for the access point/router will open, shown in Figure 13.10.

YOUR ACCESS POINT'S PASSWORD

The default password for most routers is admin (or is blank). I suggest that you leave your password at the default for now so that you won't forget it while you are setting up your network.

After your Wi-Fi network is up and running, from a security standpoint, it is a good idea to change the default password. Changing the password makes it much harder for intruders to open the administrative functions or your access point, and it is one of the easiest security measures you can take. You shouldn't delay changing it once you are sure that your network is running.

For more about Wi-Fi networks and security, see Chapter 18, "Securing Your Wi-Fi Network."

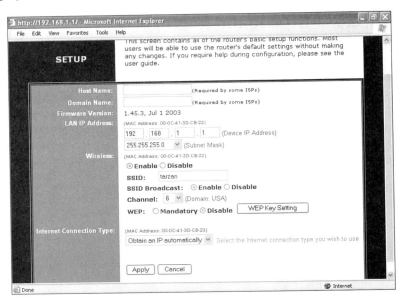

FIGURE 13.10

If you use dynamic IP addressing, your Wi-Fi network will work if you accept the default settings, but you should make a few changes for the sake of security.

Have a look at the screen shown in Figure 13.10 and the information you collected about your Internet settings. If your Internet settings use automatic IP addressing, you can probably leave the default settings as they stand and have your Wi-Fi network up and running (although I do suggest that you make a few changes for the sake of security, as I'll explain in a second).

If you use a static IP, you need to select this from the drop-down list shown at the bottom of Figure 13.10. Next, you'll need to use the data you collected to enter the IP and DNS server address information you collected in the fields that will now appear.

As I said, assuming that you use automatic IP addressing, which is most likely, you should make a few changes just to make your network more secure.

First, make sure that Wireless is set to Enable. If the Disable button is selected instead, your wireless network will not work.

Next, you should change the SSID (or network name) to something other than the default. In Figure 13.10, you can see that I changed the default name for the network (which is linksys) to something I made up, namely tarzan.

You might want to disable SSID broadcast. (The default is to enable this feature.) If SSID broadcast is enabled, the scanning feature available in most Wi-Fi cards will let them see your network name. So running disabled is a security measure because you are in stealth mode. For my own part, I tend to enable SSID broadcast because I find it useful to be able to see the SSIDs that are broadcasting when I connect a computer to one of our networks.

However you set SSID broadcast, you should certainly turn on encryption. WEP (wired equivalent privacy) protects your Wi-Fi network with a very basic level of encryption. Most equipment comes from the manufacturer with security disabled. This is done to make initial setup easier (unfortunately, it is also why there are so many unprotected access points in the world today). Make a mental note to security-enable your access point after you get it up and running.

> **TIP**
>
> WPA-PSK is an encryption setting that is somewhat more secure than WEP and easy to use on a home network. See Chapter 15 for more information about securing your wireless network with WPA-PSK.

To enable WEP, select the Mandatory option next to the WEP Setting, and click the WEP Key Setting button. The WEP Key Setting window will open, shown in Figure 13.11.

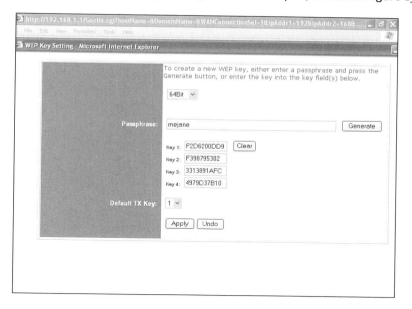

FIGURE 13.11

Your Wi-Fi network is more secure if you use a WEP key.

In the WEP Key Setting window, enter a pass phrase. (In Figure 13.11, I entered the phrase `mejane`.) Next, click Generate. The four default hexadecimal keys will be generated, as shown in Figure 13.11.

Next, click Apply to accept the key and close the WEP Key Setting window. Back in the main setup window for the access point/router, click Apply to accept the changes you've made to the settings.

> **NOTE**
>
> With most Wi-Fi cards and operating systems—such as a Centrino laptop running Windows XP—you'll just need to enter the first key to access the network. Other cards and operating systems will require all four keys.

> **TIP**
>
> Don't forget to write down the first hexadecimal key so that you'll be able to connect your Wi-Fi computers to the access point.

After the Linksys unit has been configured, you should check to make sure that you still have connectivity on the wired portion of your network by making sure that you can surf the Internet.

D-Link AirPlus Xtreme

The D-Link AirPlus Xtreme, model DI-624 has a web-based administration utility that works in much the way as the Linksys utility—but the details, of course, are different. And, as they say, the devil is in the details.

As with the Linksys unit, you should first collect your Internet settings and have a modem connected either via cable or DSL to the Internet.

Plug the D-Link unit in to the modem. Plug a computer in to the D-Link unit. After this is done, open a Web browser, such as Internet Explorer, on the connected computer.

Using the browser, open the web page http://192.168.0.1. The logon pop-up screen shown in Figure 13.12 will appear.

FIGURE 13.12

You will be prompted for a username and password when you log on to the D-Link's administrative utility.

As shown in Figure 13.12, enter **admin** for the username and leave the password field empty. Click OK.

The D-Link administrative utility will open in your browser to a page with a button that lets you run the Setup Wizard. If you click Run Wizard, the Setup Wizard will start. The initial pane of the Setup Wizard is shown in Figure 13.13.

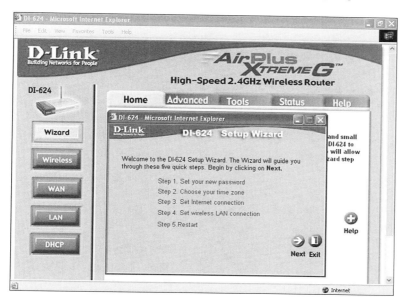

FIGURE 13.13

The D-Link unit's Setup Wizard will help configure your access point.

In Figure 13.13, you can see most of the information that the Setup Wizard gathers, and you can choose to enter it using the Setup Wizard. I've found, however, that usually the Setup Wizard doesn't gather everything needed and it's just as easy to enter your settings manually.

The first step is to enter the settings needed for your Internet connection. To do this, click the WAN button shown on the left of Figure 13.13. The WAN Settings tab, shown in Figure 13.14, will open.

Enter the information you gathered about your Internet settings on the WAN Settings tab using the following guidelines:

- If you connect to the Internet using a cable modem, you will probably select Dynamic IP Address.

- If your ISP provides a static IP, select Static IP Address. After you've made the selection, a field will appear for entering the IP.

- If you connect using DSL, you will probably need to select PPPoE. After you've made the selection, a field will appear for entering the username and password information provided by your ISP.

- In any case, you are likely to have to enter Primary and Secondary DNS addresses toward the bottom of the screen.

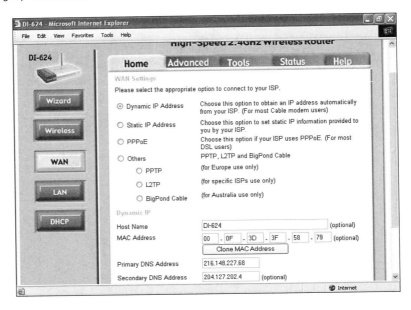

FIGURE 13.14

The WAN Settings page is used to enter the settings you need to connect to the Internet.

After all your information has been entered, click Apply (found on the lower right of the screen). Next, click the Wireless button to open the Wireless Settings tab, shown in Figure 13.15.

The Wireless Settings tab is used to configure the Wi-Fi portion of your network—that is to say, the part of your network controlled by your wireless access point.

Make sure that the wireless radio is on and provide an identifying network name (also called the SSID). Next, give your wireless network some security by enabling 128-bit WEP encryption and at least the first key. An example of these settings is shown in Figure 13.15. Click Apply when you are through.

Next, click the Tools tab to open the Administrator Settings page, shown in Figure 13.16.

You can use the Administrator Settings page to change the login password. As you'll recall, the initial password is blank. This should be changed for security reasons, but you don't have to do it right away. Before you bother with it, you can make sure that your wired and wireless networks are connecting to the Internet and that you can connect using your laptop.

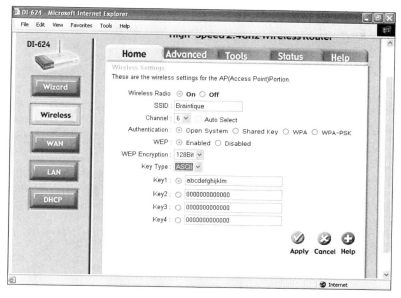

FIGURE 13.15

The Wireless Settings page is used to enter the settings you need for the Wi-Fi portion of your network.

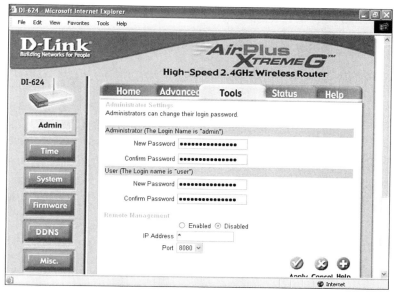

FIGURE 13.16

The Administrator Settings page is used to change the login password.

With all the settings in place in the D-Link's administrative utility, close your Web browser and reopen it and check to make sure that you can connect to the Internet.

Cloning MAC Addresses

As I mentioned earlier in this chapter, some ISPs attempt to control your access to the Internet by verifying the unique MAC identifier baked in to your network card. The MAC that belongs to your wireless access point will not match the MAC that goes with the network card in the computer you originally used to log on to the Internet, so the ISP won't let your access point connect.

If you can't figure out why you can't connect to the Internet, this might be the problem. Of course, if you think that this is the problem, you should probably contact your ISP, and tell them that you have changed your network access equipment, and get them to change the MAC address associated with your account.

You can set most Wi-Fi access points to spoof a MAC address or fool the ISP into thinking that they have the same MAC address as some other device, such as the original network card.

For example, to do this using the D-Link unit, open its administrative utility as I just explained. Next, open the WAN Settings tab shown in Figure 13.14.

As you can see in Figure 13.14, the actual MAC of the wireless unit is shown. You can manually enter a MAC to replace it, which will then be spoofed to the ISP.

If you don't know the MAC address to enter—and who walks around knowing the MAC address of the network card registered with their ISP—there are two possible approaches:

- You can find it out.

- You can tell your access point to clone it automatically.

> **NOTE**
> Other Wi-Fi access points have similar MAC address spoofing and cloning capabilities.

It's easy to find the MAC address of a network card by running the Ipconfig utility and reading the Physical Address information for the card.

To clone the MAC address of the network card of the machine connected to the D-Link access point (so that the cloned MAC address will be spoofed to the ISP), simply click the Clone button shown in Figure 13.14.

Airport Extreme

To automatically configure an Airport Extreme access point, you first need to have an Apple computer—equipped with Airport Extreme and running MAC OS X version 10—connected to the Internet. By the way, everything in this section applies equally to the

Airport Express Base Station as well as the Airport Extreme unit. But it's easy enough to administer these access points from either a Windows-based PC or an Apple computer.

The Apple computer that is Airport Extreme ready will be used to administer the Airport Extreme Base Station. The computer must be capable of connecting to the Internet using a non-Airport access method because later in the process, the Airport Setup Assistant software will use your computer's active Internet settings (the ones you can see on the Network tab of the System Preferences dialog) to configure the Airport Extreme Base Station.

With the Apple computer connected to the Internet, simply connect the Airport Extreme Base Station as explained in "Plugging in the Hardware."

Fire up the Airport Extreme Base Station by plugging it in. When the middle light of the three on the front of the Airport Extreme Base Station comes on, you are ready for the next step.

On your computer, open the Airport Setup Assistant application. (You can find it in the Applications/Utilities folder.)

The Airport Setup Assistant will first configure the Airport Extreme Base Station, using the network settings from your computer, and then configure your computer to work with the base station.

When prompted, you should provide a name for the base station (and Wi-Fi network) and a password that is not the default. By the way, it is possible that you might need to know the default password to enter these new settings; if so, it is `public`.

When you click through all the screens in the Airport Setup Assistant, your Airport Extreme Base Station and computer are good to go! See! Wasn't that easy?

Apple computers equipped with Airport Extreme will connect to the new Wi-Fi network automatically. They will prompt you for the password (encryption key) if you've protected the network—which you should do. Windows machines connect to the new Wi-Fi network the way they would to any other Wi-Fi network. (They don't have to know that Apple manufactures the Airport Extreme Base Station.) Once again, you will have to enter an encryption key to connect to a protected network.

> **NOTE** Just to keep terms straight, Airport is Apple speak for 802.11b Wi-Fi, and Airport Extreme is Apple's way of saying 802.11g.

> **NOTE** An alternative is to download the Windows administration program for Windows XP, which you can download from `http://www.apple.com/airportextreme`. With this utility, you don't need an Apple to administer the Airport Extreme Base Station.

> **TIP** If you can't find Airport Setup Assistant on your computer, use the installer for it located on the CD-ROM that came with your Airport Extreme Base Station.

> **NOTE** You can use any computer with Wi-Fi capabilities—such as a Centrino laptop—as part of the network you are creating with Airport Extreme Base Station. It doesn't have to be an Apple computer.

Connecting to the Wi-Fi Network

It's time to connect to your new network. Now that you have the devices and wires in place, and with your access point/router configured, it is time for the final step: setting up your Centrino laptop to work on the wireless network.

You'll need to know the SSID and encryption key (password) you used for the network when you configured the router.

The access point/router is now doing the job of "speaking" to your ISP and the Internet for you. It has taken over the task of assigning addresses for nodes within your internal network. So, even if you use a fixed IP within the access point/router to communicate with your ISP, the computers within your network (wireless as well as wired) should be set to automatically obtain their IP and DNS server addresses.

Open the Wireless Network Connection Properties window, shown in Figure 13.17. (See Chapter 3, "Configuring Your Mobile Computer," for information about how to work with the Wireless Connection Properties window.) Click the Add button to create a new preferred network profile. The Wireless Network Property window will open, as shown in Figure 13.18.

FIGURE 13.17

Click the Add button to create a new preferred network profile.

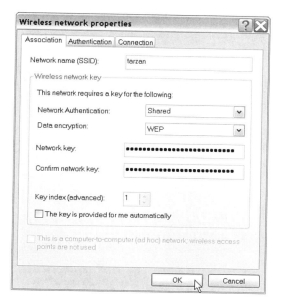

FIGURE 13.18

The Wireless Network Properties window is used to enter the encryption key for a Windows XP node on the tarzan *wireless network.*

Now you are ready to fill in the Wireless Network Properties window. To begin, make sure that the check box labeled The Key Is Provided for Me Automatically is unchecked. Next, set Network Authentication (Shared), and set Data Encryption (WEP Enabled). Finally, enter (and confirm) as the network key the first of the four hexadecimal keys you generated in the access point/router's setup program. (This was shown in Figure 13.11.) Click OK to accept the settings, and click OK in the Wireless Network Connection Properties dialog to accept the changes and close the window.

You can test that you have access to the encrypted wireless network by using your browser to open an Internet site such as http://www.google.com.

In another example, Figure 13.19 shows the Wireless Network Connections Properties window for a laptop communicating with the Braintique wireless network setup using the D-Link AirPlus Xtreme unit from earlier in this chapter.

> **TIP** If you aren't able to access the Internet, you might need to reboot your computer for the new settings to take hold.

Chapter 13 | Setting Up Your Access Point

FIGURE 13.19

The Wireless Network Connections Properties window shows the Braintique wireless network setup from earlier in this chapter.

To enter the encryption key and connect to the Braintique wireless network, select the Braintique network and click Configure. The Wireless network properties window, shown in Figure 3.20, will open.

Enter the encryption key in the Wireless network properties window just as you entered it when you configured the D-Link unit (see Figure 3.15).

> **TIP**
> You can always open the access point configuration utility to find out the encryption key you used (in case you forget it).

With the Braintique wireless network connected, you can easily check its status to get an idea of its actual performance (Figure 13.21). You should also check to make sure that you can open Internet pages, such as http://www.intel.com, and see the other computers on your internal network.

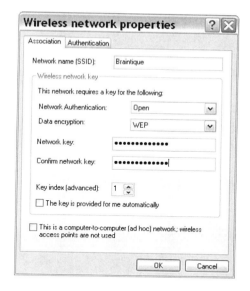

FIGURE 13.20

The Wireless Network Properties window is used to enter the Braintique network's encryption key and connect to it.

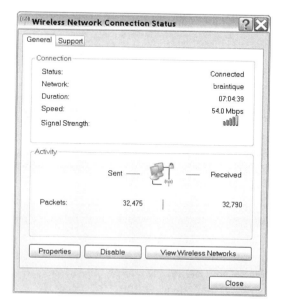

FIGURE 13.21

You can use the Wireless Network Connection Status window to gauge the performance of your new wireless network.

Setting Up a Wireless Network with Windows XP SP2

NOTE

The Wireless Network Setup Wizard makes it very easy to set up a wireless network, provided the Wizard works with your equipment. The catch is that it doesn't work with many access points, so you might still have to manually configure your wireless network.

Setting up a wireless network just got a whole lot easier with the release of Windows XP Service Pack 2 (SP2). This update to the Windows XP operating system provides a new tool for configuring wireless networks, the Wireless Network Setup Wizard.

To start the Wizard, click Setup a Wireless Network for a Home or Small Office on the Network Tasks pane of the My Network Places window.

In the Wireless Network Setup Wizard, you start by assigning a name (SSID) to the network as shown in Figure 13.22.

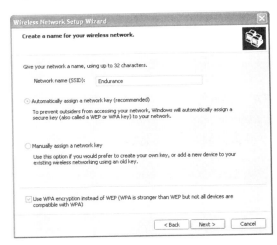

FIGURE 13.22

The Wireless Network Setup Wizard in SP2 helps you easily configure a wireless network.

On the same Wizard panel that you use to name your network, an encryption key can be chosen for you automatically, which saves you trouble, or you can enter one manually.

By default, WEP encryption is selected. Check the WPA box at the bottom of the panel to use this alternative form of encryption, explained in Chapter 15, instead. (WPA is stronger than WEP, but not all devices support it.)

Click Next to continue with the Wizard. You can then choose to proceed manually or with a USB flash drive. Manually means that the Wizard prints out the settings. You then use the printout to enter the settings in your access point and the wireless devices connected to your network.

If you've chosen the USB flash drive option, you need to have a USB flash drive connected to your computer. The Wizard then saves the wireless network settings on the USB flash drive. You take the USB flash drive to your access point and wireless computers each in turn, plug it in, and it configures your wireless network and devices for you.

> **TIP**
> If you don't see My Network Places on your Windows XP Start menu, you can open it my starting the Windows Control Panel, choosing Network Connections in the Control Panel, and selecting My Network Places from the Other Places pane.

Although most mobile computers equipped to use a wireless network probably do have a USB connection, the same thing cannot be said of access points. So bear in mind that the Wireless Connection Network Setup only works in the USB flash drive mode if your access point has a USB connection. Check to verify this before you try to use it.

Summary

Here are the key points to remember from this chapter:

- Most of the time getting a Wi-Fi network going is really easy.

- You need to have a working Internet connection with a computer connected to a broadband modem before you start.

- You must know the settings your ISP requires before you can set up your Wi-Fi network.

- The Wi-Fi access point is connected to the Internet modem, and your wired computer or network is also plugged in to the access point.

- Most Wi-Fi access points are administered using a Web browser and Web administration utility.

- You should change the default network name and enable WPA or WEP encryption for security reasons.

- After the Wi-Fi access point has been configured, you'll need to set up your computers for access to the Wi-Fi network.

- If the Wi-Fi unit doesn't work, and you think you are doing everything right, check the manufacturer's website for known service issues and consider the possibility that the unit might be defective.

Configuring Your Wi-Fi Network

In Chapter 13, "Setting Up Your Access Point," I showed you how to set up a very simple network using a Wi-Fi access point/router combination. The equipment needed for that kind of network has become quite inexpensive, and it should do for a great many home and small office applications. This chapter fills in some gaps from Chapter 13 and discusses some of the other topics involving Wi-Fi that can come up in networks. I'll start by showing you how ad hoc, or peer-to-peer, networking works with Wi-Fi. Next, I'll show you how to share your Internet connection with another computer—something you might want to do if your computers are connected using ad-hoc networking.

I'll go on to a number of topics that are involved in setting up networks and show you a number of different ways to arrange, or lay out, networks. (The arrangement of a network is called its *topology*—in case you want to use a flashy network engineering term with your friends.)

Finally, I'll finish the chapter by showing you a good way to set up a Wi-Fi hotspot—in case you ever decide to share your wireless network with the world.

Ad Hoc Networking

Ad hoc networking means that each computer talks to each other directly without the "supervision" of a device such as a router. This arrangement is sometimes called peer-to-peer networking. In wireless networking, a strength (and sometimes weakness) of ad hoc networking is that nodes on the network (meaning computers) can join (or leave) the network on-the-fly. So you don't necessarily know the number of computers on a wireless network.

Figure 14.1 shows an example of an ad hoc network that uses wireless networking.

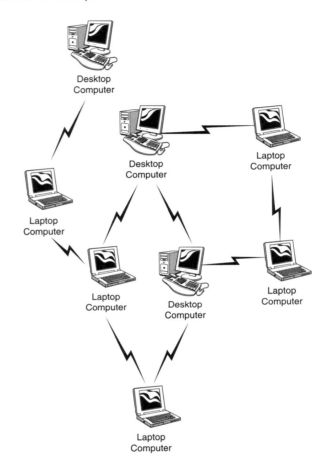

FIGURE 14.1

An ad hoc wireless network.

If enough computers are involved, an ad hoc, or peer-to-peer network, can begin to form a kind of grid, or mesh. This gives peer-to-peer networks in some applications a great deal of power, although there's no really good way to administer a peer-to-peer network, and security remains an issue.

The Wi-Fi standards specify two different configuration modes, infrastructure and ad hoc. The access point/router style network that I showed you in Chapter 13 is an example of using Wi-Fi in infrastructure mode, whereas (as you'd probably expect) peer-to-peer Wi-Fi access uses ad hoc mode.

In infrastructure mode, communications between two nodes on the network flows through the access point. Computer A actually "talks" to the access point, which in turn talks to computer B. The access point also performs a number of other roles, such as connecting the nodes to the Internet or other WAN (wide area network), connecting multiple wireless networks, connecting the wireless nodes to a wired network, and providing management and security functionality (such as a firewall, as explained in Chapter 17, "Protecting Your Mobile Wi-Fi Computer," and Chapter 18, "Securing Your Wi-Fi Network").

In contrast, in ad hoc mode, computers A and B communicate directly without an intermediary, and none of the other functionality provided by the access point is present.

One drawback of using Wi-Fi in ad hoc mode can be signal strength and range. Ad hoc mode might make sense to use if you had two computers close to one another that you didn't expect to move much—an advantage being that you wouldn't need to spend $50 on an access point.

Another drawback of ad hoc mode is transfer rate. The IEEE 802.11 specification states that when in ad hoc mode, products only need to support a transfer rate of 11Mbps. In some cases products may exceed the specification and run at the 54Mbps rate of 802.11g but this is not required of the manufacturers so it can't be guaranteed.

I wouldn't maintain a fixed Wi-Fi network using ad hoc connections with more than a few computers, and I wouldn't expect to be able to maintain an ad hoc connection if one of my computers were mobile. When my wife carries our Wi-Fi laptop out to the garden, it goes way out of ad hoc range, but well within the broadcast range of our access points. (Nothing inherently causes an ad hoc broadcast to have less range; it's just that usually the computers used for ad hoc broadcasts don't have as good a radio or antenna and don't transmit with as much power, as a dedicated access point.)

Ad hoc connectivity is good for, well, ad hoc situations in life. Using Wi-Fi's ad hoc mode, you can connect two computers when you are on the road even though no wireless networks are present, which can be very useful. For example, on a road trip, a colleague might have a wired network connection and access to the Internet, which I might be able to share using an ad hoc connection with my colleague. (Sharing an Internet connection is explained later in this chapter.)

Ad hoc networks are also good for quickly and easily setting up a Wi-Fi network in situations in which flexibility is essential and a wireless infrastructure is not available (or needed). An example of this might be a temporary meeting or convention, or perhaps a group of consultants working onsite but without access to the "official" network might decide to set up an ad hoc network to exchange files between themselves.

Later in this chapter, I'll also show you a hybrid form of networking, which uses a Wi-Fi access point and an ad hoc mesh to blanket a large area and extend the range of the core configuration.

Setting Up Ad Hoc Wireless Networking

The process of establishing an ad hoc network involves two steps:

1. Setting up the network on one machine

2. Connecting to the ad hoc network from one or more other machines

In this section, I'll show you how to set up an ad hoc network between two laptops with Intel Centrino mobile technology running Windows XP. We'll start by defining the ad hoc network on your computer, then we'll go to your friend's computer and join it. Neither step—setting up the ad hoc network and connecting to it—is difficult. In particular, connecting to an ad hoc network is pretty much the same as connecting to any other wireless network.

Starting at your computer, open the Network Connections window either by right-clicking on the wireless connection icon in the taskbar and selecting Open Network Connections, or by selecting the Network Connections item from the Windows Control Panel. With the Network Connections window, select Wireless Network Connection, right-click, and choose Properties from the context menu. The Wireless Network Connection Properties window will open. Click the Wireless Networks tab and you should see a dialog similar to Figure 14.2. (See Chapter 3, "Configuring Your Mobile Computer," for more information about the various Network Connection windows and how to use them.)

FIGURE 14.2

The Wireless Network Connections Properties window is the starting place for creating an ad hoc network.

Click the Advanced button at the bottom of the Preferred Networks section. This will display the Advanced dialog and give you three choices of networks to access. Click the third radio button, Computer-to-Computer (Ad Hoc) Networks Only. Also make sure the Automatically Connect to Non-Preferred Networks check box is unchecked. The dialog should now look exactly like Figure 14.3. Click the Close button. Doing this step now makes the definition of your ad hoc network a little easier and helps prepare your computer to automatically enable ad hoc networking mode.

Now you should be back to the Wireless Networks tab of the Wireless Network Connection properties dialog. Click the Add button. The Wireless Network Properties window, shown in Figure 14.4, will open. This is where you define the ad hoc network.

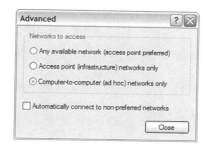

Figure 14.3

Set the Advanced dialog to access only ad hoc networks.

In the Wireless Network properties window, enter a name for the ad hoc network (the wireless network's SSID). (In Figure 14.4, I've given the ad hoc network the name theHoc.) Note that the check box This Is a Computer-to-Computer (Ad Hoc) Network; Wireless Access Points Are Not Used at the bottom of the window is already selected and greyed out.

Select the Data Encryption option from the drop-down list if you want to provide WEP encryption protection for the ad hoc network.

Click OK to save the ad hoc network profile you just made. The Wireless Network Connection Properties dialog should look like Figure 14.5.

Finally, click OK on the Wireless Network Connection Properties dialog to make your ad hoc profile active.

Your computer is now sending out an 802.11 signal called a beacon to announce an ad hoc network is available and open for

NOTE In this example I have elected not to use encryption to get the ad hoc connection up and running as quickly and easily as possible. If your ad hoc network is going to be running for only a few minutes you might choose to leave encryption off. On the other hand, if you expect your network to be up for any length of time, and especially if there are other people around, you should definitely enable encryption after you have your initial connection established and confirmed your ad hoc network is operational. With encryption off, there is nothing to prevent someone else from detecting and joining your ad hoc network. The files you exchange with a friend or colleague are just as visable to an outsider. Ad hoc networks are easy to set up and join, so make sure this benefit is not turned against you.

business. There is nothing to tell you this is happening on your computer—there is no status message or icon to look at. You just have to go on faith that your computer is ready and waiting for another computer to connect to it.

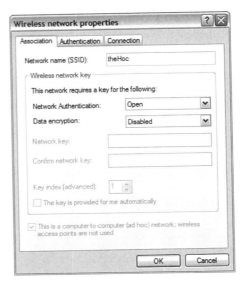

Figure 14.4

The Wireless Networks Properties window is used to define a wireless network for ad hoc broadcast.

Connecting to the Ad Hoc Network

Your Centrino mobile computer is now happily broadcasting to the world and an ad hoc network is available and ready. Now it's time to connect your friend or colleague to the new ad hoc network we defined in the previous section, "Setting Up Ad Hoc Wireless Networking." (You'll recall that I named the network theHoc.)

The steps are very much the same as defining the ad hoc network in the previous section, but this time we will have the satisfaction of watching the two computers connect to each other, then testing the connection out. OK, here we go.

Using your friend's mobile computer, open the Wireless Network Connections properties window, and click on the Wireless Networks tab. You should have a dialog that looks similar to Figure 14.2. (See the previous section, and Chapter 3, for more information about working with the Wireless Network Connections properties window.)

FIGURE 14.5

The new ad hoc network appears in the Wireless Network Connection properties window.

Just as we did with your computer, click the Advanced button in the Preferred Networks section and select the third radio button, Computer-to-Computer (Ad Hoc) Networks Only. Once again, this dialog should look exactly like Figure 14.3. Be sure the check box Automatically Connect To Non-Preferred Networks is unchecked, and then click the Close button.

Now you should be back at the Wireless Networks tab of the Wireless Network Connection Properties dialog. Click the Add button. The Wireless Network Properties window, shown in Figure 14.4, will open. This is where you define the ad hoc network on your friend's computer so the two can connect to each other. Just as you did before, enter the Network Name (SSID) theHoc and click OK.

Now for the moment of truth. The Wireless Network Connection Properties dialog should still be displayed with the Wireless Networks tab selected, just as in Figure 14.5. Click OK. As soon as you do this, both computers will recognize each other and start to negotiate the connection. The wireless connection icon in the taskbar should change from showing a red X to a small dot moving back and forth. Within 30 to 60 seconds you should see the connection message shown in Figure 14.6.

NOTE

It is extremely important that the Network Name (SSID) you enter on each computer in your ad hoc network be exactly the same. A simple typo will prevent the connection from being established. And yes, case matters, so make sure your upper- and lowercase letters are the same, too. Once it's created, the SSID cannot be changed in the profile. If you make a mistake, you will need to delete the ad hoc profile from the Preferred Networks list (use the Remove button), and then create a new one with the correct name.

FIGURE 14.6

The balloon message in the taskbar indicates a successful connection.

That's it! Your computers are now connected without wires. You can test your connection by doing a simple Windows networking operation. Figure 14.7 shows how I was able to browse files in the shared document folder on a laptop called Nomad from another computer connected to the ad hoc network.

FIGURE 14.7

You can test your ad hoc connection by viewing the shared documents folder on another computer.

Connecting a third, fourth, or tenth computer to the ad hoc network is achieved in the same way shown here in this section. With two or more computers communicating with each other, your ad hoc network should even be visible in the available networks window as shown in Figure 14.8. Note the different icon used to distinguish among available infrastructure (access point) and ad hoc networks.

One last thing. In the process of setting up the ad hoc network, we configured your computer to connect only to wireless ad hoc networks. This could be a problem when you return home or arrive back at the office and want to connect to the wireless network there. Go back to the Advanced settings menu for preferred networks (shown in Figure 14.9) and click the top radio button labeled Any Available Network (Access Point Preferred) to return your system to the normal default setting.

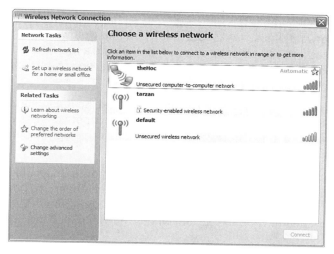

FIGURE 14.8

Ad hoc and infrastructure networks (access points) show up differently in the View Available Wireless Networks window.

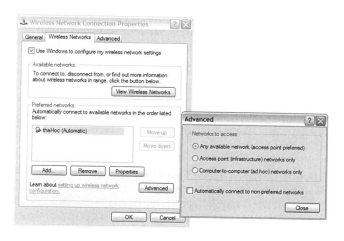

FIGURE 14.9

Remember to reset your system to allow connection to access points when you are finished with ad hoc networking.

When Ad Hoc Networking Has Problems

In the previous example I took you through the setup of an ad hoc network from beginning to end where there were no problems. And with a modern Centrino-based mobile computer you should have the same experience I had in writing up this example—easy and straightforward, it just works. But you know, and I know, that sometimes the unexpected happens and you have to cope with it. Here are a few tips and tricks to help you move beyond some problems in getting computers talking to each other in an ad hoc network.

Be forewarned that this information is fairly technical. So if reading about the inner workings of 802.11 protocols and TCP/IP addressing is not your cup of tea, you can skip this section, or get a geeky friend to read it and she can help you sort out any problems you might have with an ad hoc network.

If you are trying to connect your Centrino mobile computer to an older computer running an earlier version of Windows (98, Me, or 2000) in theory ad hoc networking should work, but there are some issues you need to know about. Even an older computer upgraded to Windows XP, but with an early model 802.11 card can have difficulties.

The 802.11 standard for ad hoc networking evolved from a more primitive form to the implementation in use now. The early kind of ad hoc networking (called ad hoc demo mode or AHDM) is not interoperable with the final standard for ad hoc networking called IBSS (for independent basic service set).

I was unfortunate enough to have an older 802.11 card and wasted four hours trying to make the darn thing work only to finally discover it was using the old AHDM set of ad hoc protocols. Lesson learned: If the ad hoc network does not come up immediately, and you have an older computer with an early 802.11 interface, give it up. Don't waste your time.

One of the enormous advantages of a new Centrino-based laptop is the maturity of the technology. All the kinks and quirks have been ironed out and the products just work. You won't waste your time fiddling with this and that, trying to coax it into behaving nicely.

Not all problems with ad hoc networking are as severe as the interoperability problem between AHDM and IBSS. Here are a couple tricks you can try if you are having trouble getting your Centrino-based mobile computer to communicate in an ad hoc network with an older computer (or even a new one that is not Centrino based).

Some 802.11 wireless interfaces are capable of setting special properties that affect how the interface functions. Figure 14.10 shows some of the settings for the Intel PRO/Wireless 2200BG Network Connection. In particular note the setting called Ad Hoc Channel that is set to a value of 11. This setting tells the interface which of the 11 or so

802.11 channels to broadcast on when advertising an ad hoc network. It is also the first channel this interface will use when trying to locate an ad hoc network. It is not necessary to change this setting to get Centrino-based laptops talking to each other because they scan all available channels to find ad hoc networks. But that might not be the case with an older interface. So one tip is to make the ad hoc channel setting on all the wireless adapters the same.

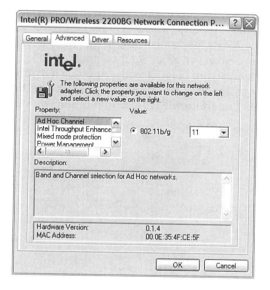

FIGURE 14.10

Setting the ad hoc channel setting the same on all participating computers might help solve ad hoc interoperability problems.

The final piece of advice has to do with TCP/IP network addressing. Most computers today get their numeric TCP/IP network address from a dynamic host configuration protocol (DHCP) server. Whether your computer connects to a wired network or a wireless one, the first thing it does is send out a message requesting an address. The process works so well, almost no one knows or cares to change TCP/IP address settings anymore.

In an ad hoc network there is no DHCP server to give an address. Your Centrino-based Windows XP mobile computer is still going to try to get an address when it connects to an ad hoc network, but after about 60 seconds or so when it doesn't hear back from a DHCP server it gives up and Windows Automatic Private IP Addressing (APIPA) takes over. Windows XP generates its own semi-random address of 169.254.xxx.xxx, where the xxx represents a number picked at random between 1 and 254.

Windows XP systems work just fine with the APIPA-generated addresses. You can refer to the computers by their human readable names (Nomad instead of 169.254.252.12), browse files, and even print to a printer if one is attached. But not all computers are as forgiving as Windows XP. So for these less flexible systems you will need to assign a consistent set of TCP/IP addresses in the same subnet.

Windows XP has an extremely nice facility that enables you to have your cake and eat it too when it comes to network addressing. Figure 14.11 shows the Alternate Configuration tab on the Internet Protocol (TCP/IP) Properties dialog. You can use this setting to override the APIPA mechanism and specify any address you want when no DHCP server is present. However, when your computer does get a response from a DHCP server (like when you are on your home or office network) it will prefer that and take the dynamically assigned address instead. This way, setting up addressing for an ad hoc network doesn't interfere with the configuration you want for your home or office.

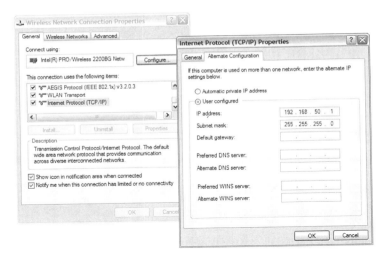

FIGURE 14.11

You can configure Windows to use a specific IP address when no DHCP server is present.

It is a TCP/IP addressing convention that IP addresses of 192.168.xxx.xxx are for private networks. You can see that I specified 192.168.50.1 for the computer in Figure 14.11. So following this example I would start at 192.168.50.1 and give each computer its own address incrementing by one for each system (192.168.50.1, 192.168.50.2, 192.168.50.3, and so on).

If you are going to shut off APIPA and assign your own addresses, you should be aware this is an all-or-nothing proposition. Either all the systems in your ad hoc network are Windows machines that use APIPA-generated addresses, or all of them use IP addresses that you manually assigned. It's not possible to have a mixture of some systems with APIPA and some with manually assigned addresses (not if you want them all to communicate with each other).

Sharing a Connection Through a Computer

It's sometimes very useful to share your computer's existing Internet connection with another computer, particularly when you are using ad hoc wireless networking.

For example, consider the following scenario: You and a colleague are working to put together a presentation for an important meeting. You have traveled to a strange city to see an important client. Your hotel room has an Ethernet wired connection to the Internet. But your mobile computer does not have a network interface card. Fortunately, your colleague's mobile computer does—and both of you are using Intel Centrino laptops.

Your colleague connects to the Internet and creates an ad hoc wireless network. He shares his connection to the Internet. You connect to his improvised ad hoc wireless network and use his shared connection to the Internet to gather the information you need to create a winning presentation.

It's easy to share an Internet connection from a mobile computer with Intel Centrino mobile technology.

Open the Control Panel from the Windows XP Start menu. In the Control Panel, double-click Network Connections.

With the Network Connections window open, highlight the connection to the Internet you want to share. With the connection highlighted, open the Properties dialog for the connection either by right-clicking the connection and selecting Properties from the context menu or by choosing the Change Settings of This Connection option under Network Tasks on the Network Connections window.

> **NOTE**
> Sharing, as I tell my kids, is good. And sharing an Internet connection can be very useful. But it does pose some security concerns. So if you are not actively sharing your Internet connection, you should run with sharing turned off as a default.

With the Properties dialog for the connection open, click the Advanced tab. The Advanced tab for the Properties dialog is shown in Figure 14.12.

To enable Internet connection sharing, check the Allow Other Network Users to Connect Through This Computer's Internet Connection box.

Click OK. Your Internet connection is now shared.

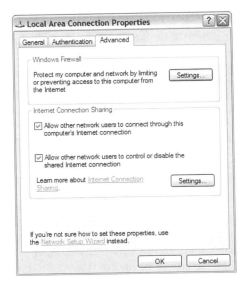

FIGURE 14.12

The Advanced tab of the Properties dialog for the connection is used to enable or disable Internet connection sharing.

To reverse the process, and disable Internet connection sharing, simply clear the Allow Other Network Users to Connect Through This Computer's Internet Connection check box.

Troubleshooting a Network

Most of the time, if you are deploying a simple Wi-Fi network such as the infrastructure style network described in Chapter 13 or a one-to-one ad hoc network such as the one explained so far in this chapter, you won't have any problems setting up a network. But sometimes networks go bad.

WHAT DOES IT MEAN FOR A NETWORK TO GO BAD?

You know that you are having network problems when... one or more nodes on the networks stop being capable of "seeing" other nodes on the network, or some (or all) of the nodes on the network stop being capable of accessing the Internet.

Although many network problems are easy to find and fix—for example, an unplugged cable—others can be more subtle.

It can be very difficult to diagnose network problems (although fixing the problem when you find it can be fairly straightforward). The first thing you should suspect in a wireless network is that your node is out of range of the access point. You should also check the connections in the wired part of a network. I've wasted much time in fruitless network debugging when the problem all along was a network cable that had come unplugged. Defective network cables are another problem. (I'm mentioning cables and wires here because in a mixed wire line-wireless network, problems with cabling can cause problems in the Wi-Fi part of the network.)

If your're having trouble with a network and can't identify the cause, it's always worthwhile turning *everything* off and on—and seeing if the problem goes away!

Start by turning off your broadband modem. If it doesn't have an on/off switch, you might have to pull the plug. Next, turn off your router and access point. Once again, if your devices don't have power switches, you should simply pull the plug. Finally, turn all your network computers off.

Wait at least a full minute. Then turn everything back on in the same order you turned them off.

> **TIP** If a network cable has been crimped or damaged in any way—for example, the casing was broken by a staple when you were positioning cables—you should not take any chances, but just throw the network cable out and start over with another cable.

A surprising amount of times, I've found that this fixes the problem—which should probably then be attributed to some cosmic force that has "gunked up" the network or the gateway to the Internet.

Troubleshooting with Ipconfig

Another possible cause of networking problems has to do with the way IP addresses (the way nodes are identified on the network) are being assigned. Perhaps IP assignment has gone wrong so that two devices on your local network have been assigned the same IP. In this case, the Ipconfig program is a very valuable tool.

If you are trying to diagnose and fix a network problem, you should know about the Ipconfig program as one of the troubleshooting arrows in your quiver.

Ipconfig can be used to display TCP/IP network configuration values, discard the current IP and DHCP settings for a device, and renew the DHCP settings for a device. (For more about working with DHCP, see Chapter 15, "Advanced Access Point Configuration.")

If your computer is connecting to the Internet or your local network properly, an easy thing to try is to use Ipconfig to release (meaning discard) its current settings and then renew itself with new settings.

Ipconfig is a command-line program. To see the results of the program, you should run it from a Command Prompt box. To open a Command Prompt box, select the Command Prompt icon in the Accessories program group from the Windows XP Start menu.

Next, at the command line, type the command you want to run. Here are some of the most important ways you can use Ipconfig:

The command Ipconfig /? displays all the Ipconfig commands and the syntax of the program. So this is the command to run if you want to learn more about what Ipconfig can do and how to use Ipconfig.

The command Ipconfig /all displays the network settings for a TCP/IP device on the network, as you can see in Figure 14.13. You can use this information to track the IP addresses assigned to computers on your network and make sure that there is no conflict because two computers have been assigned the same address. You can also use the IP address of a device on the network to access the device directly without knowing its name.

FIGURE 14.13

Ipconfig /all *shows the network settings for a computer on the network.*

The command Ipconfig /release sends a message to the DHCP server to release the current IP address for a device on the network.

The command Ipconfig /Renew sends a message to the DHCP server to renew the IP address of your computer, provided your computer is set up to automatically obtain its IP address. The results of running this command on my computer are shown in Figure 14.14.

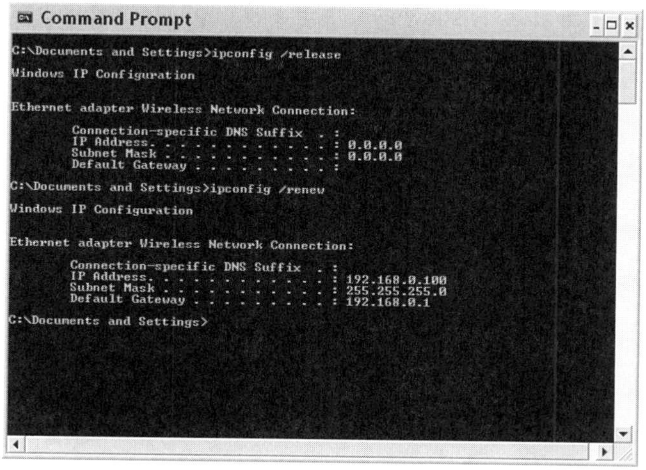

FIGURE 14.14

`Ipconfig /renew` *assigns a new IP address to a computer that obtains its IP addresses automatically.*

Understanding Networks and Their Layouts

Your office or home network might be small right now, but it might grow over time (which is what happened to me). This section explains some of the ways to look at larger, more complicated networks. This material is somewhat advanced and can be something to keep in the back of your mind when you are setting up your first Wi-Fi access point. But you might find that it provides a useful perspective even as you are getting started. The point of this section is to let you know what your options are as you start creating a wireless, or mixed wired and wireless network. After all, by the time you add computers for all your kids, an at home office, and wireless streaming media, you'll find that your network isn't so small anymore. So you don't have to worry about the material in this section when you are building your first wireless network with one or two computers. But you might want to think about it as your networks get more complex.

The way that a network is laid out, or arranged, is called the network's *topology*. The term topology also refers to how the devices on the network communicate with each other. A network's *physical* topology is the way devices are laid out (meaning which devices are connected to each other, and so on). In contrast, a network's *logical* topology is the way that the signals act on the network media, meaning the way that the data passes through the network from one device to the next independent of the physical interconnection of the devices. If you diagrammed the physical and logical topology for a network, the diagram might look the same; but then again, it might not.

I've already shown you some examples of network topologies. For example, the mesh pattern created by an ad hoc wireless network used by a number of devices is a network topology.

An infrastructure wireless network that uses an access point is operating in a star topology, as shown in the diagram in Figure 14.15.

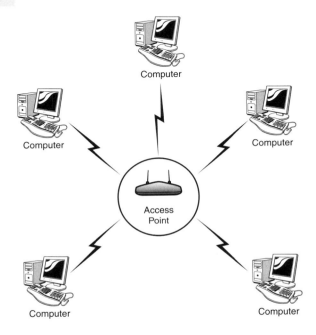

FIGURE 14.15

An infrastructure Wi-Fi network that uses an access point is an example of a star topology.

Two other common kinds of network topologies are the bus (see Figure 14.16) and the ring topologies, shown in Figure 14.17. In the bus topology, all the devices on the network are connected to a central cable or bus (also called backbone).

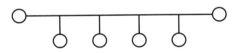

FIGURE 14.16

In a bus topology, all devices are connected to a central cable, whereas in a ring topology, the devices are arranged in a loop so that each device communicates with the two devices next to it.

In a ring topology, all the devices are connected to one another in the shape of a closed loop so that each device is directly connected to the two devices on either side of it, and only those two devices. (Does the ring topology "rule them all"?)

> **NOTE** Networks based on a star topology are probably easier to set up and manage than any other kind of network.

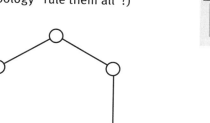

FIGURE 14.17

In a ring arrangement, the devices are arranged in a loop so that each device communicates with the two devices next to it.

As a practical matter, most even reasonably complex networks are hybrids that have features of a variety of topologies. For example, one common hybrid is called the *tree* topology, in which groups of star topology networks are placed along a bus-topology backbone. This is a common arrangement for Wi-Fi networks that cover a large area. Each Wi-Fi access point manages a star topology group, and the Wi-Fi access points are connected using a backbone.

From the viewpoint of Wi-Fi, another interesting hybrid topology involves creating a full mesh network. Yes, as I explained at the beginning of the chapter, you can create a kind of mesh using Wi-Fi devices in ad hoc mode. But what you have isn't necessarily very reliable, and it also doesn't facilitate full mesh computing in which every device speaks to every other device. (The devices don't necessarily have to speak to each other at the same time, or in real time.)

A useful Wi-Fi mesh network needs to be capable of responding to devices entering and leaving the network, operating with devices in infrastructure mode as well as ad hoc mode, and switching access points into bridges as required. Figure 14.18 shows what a mesh network of this sort might look like.

A true Wi-Fi mesh network of the sort shown in Figure 14.18 has a great many advantages. It has no single point of failure and is therefore self-healing. This kind of network can easily get around obstacles—such as water-laden foliage and barriers to sightlines—that are problematic for other kinds of Wi-Fi networks.

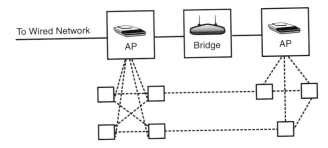

To Wired Network

FIGURE 14.18

A true Wi-Fi mesh network has a great many advantages.

However, to work properly, a Wi-Fi mesh topology requires some specialized software that provides routing functionality and has the capability to operate in both infrastructure and ad hoc modes as required.

Because of the benefits of Wi-Fi mesh networking, a lot of work is going on right now in this area. For an effort to bring Wi-Fi meshes to the third world using open source technology, you might be interested in the Wireless Roadshow project, `http://thewirelessroadshow.org`. Closer, perhaps, to home, a startup company named Firetide, Inc.—`http://www.firetide.com`, based in Hawaii—is making the hardware and software needed to deploy robust true mesh topology Wi-Fi networks. And Mesh Networks, `http://www.meshnetworks.com`, claims to be the leading provider of high performance ad hoc networking products, including turnkey mesh network solutions.

Setting Up a Hotspot with a DMZ

> **NOTE**
> If you are setting up a commercial hotspot, you should get the advice of the Wi-Fi network provider you will be working with in planning the hotspot (unless you expect to be doing service provisioning yourself). You should also know that a number of turnkey "put up a hotspot" kits are available, which you can buy and not have to think about further.

Suppose that you have a small office with a network and want to set up a public Wi-Fi hotspot. The single most important requirement is that people who use the Wi-Fi hotspot should *not* be able to access the office network. This is important because it maintains the integrity (and security) of the office network. It's easy to imagine needing this in real life, even if you are not out to compete with Starbucks.

For example, suppose that you run a small engineering business that sometimes uses part-time consultants. The consultants use your home office on occasion and need access to the Internet for their email and other uses. But you don't want to give them access to the resources in your private network, including the personal documents on your computers.

There are many ways to set up a network to provide Internet access without also providing access to your personal network. Which method to use depends on the precise functionality required and the level of security needed.

The key concept used to protect the private network is the *DMZ*. DMZ is a term borrowed from the military that is short for *demilitarized zone*. In networking terms, it means a computer or subnetwork sitting between an internal network that needs to remain secure and an area that allows external access—for example, a Web server or a Wi-Fi hotspot.

Figure 14.19 shows a simple model of a DMZ that uses firewalls to protect the private network both from the Internet and from users of the public Wi-Fi hotspot.

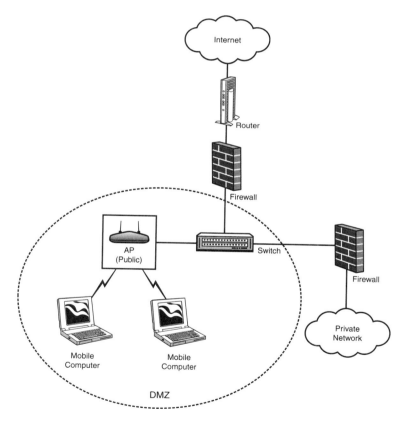

FIGURE 14.19

You can use a DMZ to protect a private network from users who have access to the public hotspot connected to the network.

The beauty of this network topology is that anyone can use the access point and wireless Internet connectivity without you having to worry about the safety or security of your private network.

If you think that you need to set up a DMZ, perhaps because you will be allowing users access to your Wi-Fi Internet access, but want to make sure that they cannot put your private network at risk, start by installing a firewall for your network, as I explain in Chapter 18. Next, read the firewall vendor's documentation to understand how to best go about setting up your firewall.

Summary

Here are the key points to remember from this chapter:

- It's easy to create ad hoc Wi-Fi connections between Wi-Fi–equipped computers.
- Sharing an Internet connection is easy and useful with ad hoc connections, but it can pose a security risk.
- Ipconfig can help you resolve Internet configuration issues with your computers on your network.
- Simple star-topology access point networks are easy to set up and manage, but when your networks start getting complicated, there are many things to think about.
- There are many ways to arrange a network.
- If you need to provide public access to your wireless network, you should consider creating a network that defends the integrity of your private resources.

Advanced Access Point Configuration

This chapter explains in detail how to set up your access point—so that this device, which is the brains of your wireless network, can do its job the way that *you* want it to.

It turns out that the access point used to control your network is a very powerful device. Assuming that you are using a combination wireless access point and router such as the units described in Chapter 12, "Buying a Wi-Fi Access Point or Router," you have a great deal of control over both how the computers on your network interoperate (and "talk" to the access point) and how your network accesses the Internet.

In this chapter, I'll start by providing some additional information you might need to connect to the Internet and to keep your access point up-to-date. Next, I'll talk about access control mechanisms built in to your access point that relate to your LAN (local area network). I'll also explain some mechanisms built in to your access point that allow you to control the access that computers on your network have to the Internet.

It's a good idea to know something about the basic technologies that make the connection work between the Internet and your home or small office network. In this chapter, I'll explain static versus dynamic IP addressing, Dynamic Host Configuration Protocol (DHCP), Network Address Translation (NAT), and DNS (Domain Name Service).

Finally, I'll show you a good alternative way to set up a mixed wired-wireless network that differs from the scenario I explained in Chapter 13, "Setting Up Your Access Point."

Dynamic IP Addressing Versus Fixed IP Assignment

If your Internet connection works using a dynamic IP (Internet Protocol) address, it means that every time your computer starts up, it is assigned an IP address from the pool of available addresses. The same thing is true when your IP is released and renewed using Ipconfig. In other words, you don't have a fixed IP address on the Internet. You can think of dynamic IP addressing as similar to time-sharing at a holiday resort: You can pack more travelers in because nobody owns anything specific, but you don't know for sure exactly who will be at a specific location at any given time.

Dynamic IP addressing is the most common way ISPs such as your cable or DSL provider use to assign you an IP address on the Internet. The IP address that the ISP dynamically provides you is used by your computer to identify itself on the Internet. It's also used by an access point/router to identify your whole network to the Internet. (The router also assigns IP addresses within your network, but that's a different story.)

But there are some things you can't do with a dynamic IP. For example, in order to host a website from your computer (as opposed to a site that an ISP hosts), you need to have a fixed (or static) IP address. You need a "location," meaning Internet IP address, at which your website can always be found.

So for this reason, and for other reasons known only to themselves, some ISPs provide you with one or more fixed IPs, rather than dynamically generated IPs.

If you have a fixed IP, you will need to enter it in your access point/router, or it will not be capable of accessing the Internet.

You can ask your ISP about the IP settings you need. Alternatively, if you are running Windows XP and are connected to the Internet, you can read the fixed IP settings from the Internet Protocol (TCP/IP) Properties dialog, as shown in Figure 15.1.

> **NOTE**
> Generally, if you are setting a static IP address for your network, you'll also need to provide an address for a DNS server. DNS is further explained later in this chapter.

> **NOTE**
> If your ISP uses dynamic, rather than fixed, IP addressing, then the IP settings will be empty (there is nothing to read).

> **TIP**
> For an explanation about how to open your Internet Protocol (TCP/IP) Properties dialog, see Chapter 13.

STATIC ON THE OUTSIDE—DYNAMIC INSIDE!

It's really important to understand that even if your ISP uses a fixed IP for your connection to the Internet, the computers on your LAN (the local area network in your home or office) most likely use dynamic IP addresses. These dynamic IP addresses are assigned by the access point.

So after the access point router has been loaded with your fixed IP settings, you need to set each computer on your network to accept dynamic IP addressing. You can think of the access point/router as acting on behalf of all the computers on your network. It connects to the network using whatever settings—fixed or dynamic IPs—and then, on its own initiative, it assigns IP addresses to your computers within your network.

In some cases, you might want to assign a static IP address to a computer on your private network. This computer then has a fixed IP address, even though the other devices on your network have addresses assigned dynamically. One reason you could assign a fixed IP address to a computer is to prevent it from accessing the Internet. For example, suppose that you want your kids to be able to share files on your network, but not to surf the Internet. For more information, see "IP Filtering" later in this chapter.

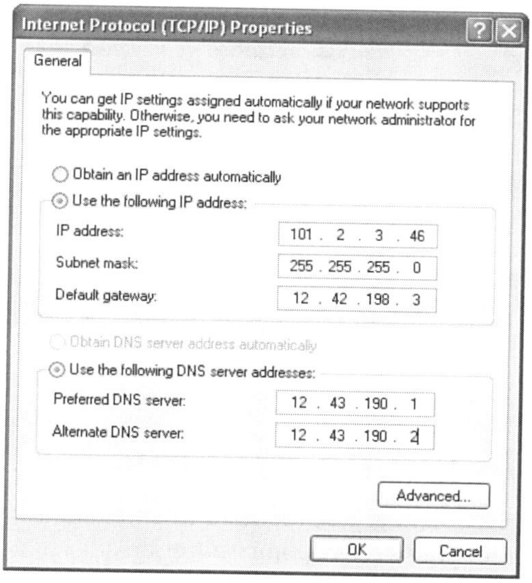

FIGURE 15.1

You can read your fixed IP and DNS settings off your computer connected to the Internet using the Internet Protocol (TCP/IP) Properties dialog.

Now that you've got the IP (and DNS) settings required by your network, you can open the administrative application for the access point/router. Figure 15.2 shows configuring the Linksys Wireless Broadband Router by selecting Static IP as the Internet Connection type, and then entering the information you gathered in the previous section. Click Apply to enter these new settings in the access point/router.

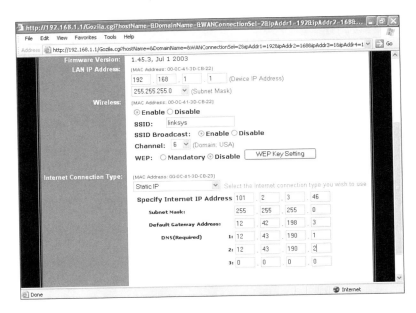

FIGURE 15.2

The static IP settings are entered in the access point/router, using the device's administration panel.

Settings That Might Be Required by Your ISP

Depending on your ISP, you might be required to enter some other settings using the administration panel for your access point/router besides just selecting dynamic IP addressing (or providing a static IP and DNS server address).

Hostname and Domain Name Settings

Some ISPs (particularly cable operators) require a hostname and domain name. Figure 15.3 shows these being entered in the Linksys Wireless Broadband Router administration panel.

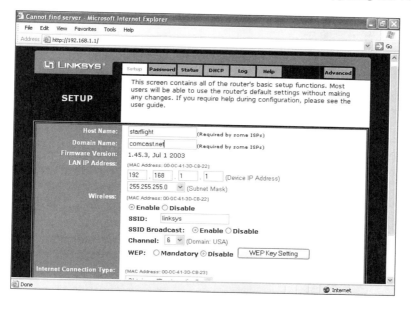

FIGURE 15.3

Some cable ISPs require a hostname and a domain name, which can be entered in the access point/router using the device's administration panel.

Click Apply at the bottom of the administration panel to keep these settings.

Point-to-Point Protocol over Ethernet

Many DSL providers use PPPoE (Point-to-Point Protocol over Ethernet). If your ISP uses PPPoE, you need to select PPPoE from the Internet Connection Type drop-down list in your access point/router's administration panel. Figure 15.4 shows PPPoE selected in the Linksys Wireless Broadband Router.

With PPPoE selected as the connection type, you'll need to enter your username and password, as shown in Figure 15.4. Click Apply to save the settings.

Upgrading Your Firmware

Things change! There's never any doubt about that, particularly in the world of technology. Standards move on; bugs are found; methods are improved.

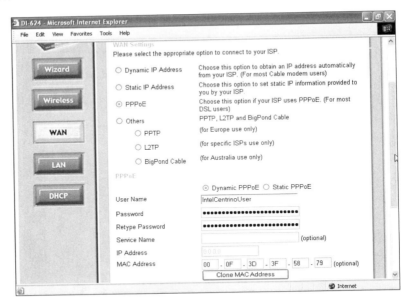

FIGURE 15.4

Many DSL ISPs use PPPoE. If yours does, you need to select it as the connection type.

It's good to know that if there's reason to change something important about the logic embedded in your wireless access point, you generally can. Updating this logic is done via the process known as a firmware upgrade.

Firmware for an access point is generally updated using the administrative application for the access point. The details differ depending on the access point and its administrative program.

> **TIP**
> You should particularly keep your eyes open for firmware updates if you have bought a wireless access point that uses a "bleeding edge" standard in which all the details have not been resolved.

For example, to update the firmware associated with the D-Link AirPlus Xtreme Wireless Router, with your unit connected to the Internet, open the D-Link's administrative program. Click on Tools to open the Administrator Settings panel shown in Figure 15.5.

Next, click Firmware. The Firmware Upgrade panel, shown in Figure 15.6, will open.

Make note of the current firmware version for your access point (shown in the Firmware Upgrade panel in Figure 15.6). Next, follow the Click Here to Check for an Upgrade on our Support Site link.

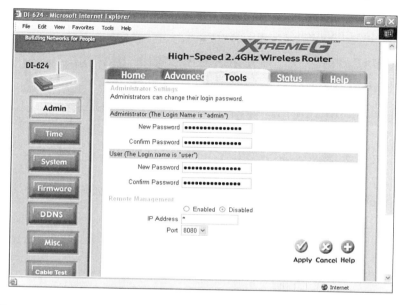

FIGURE 15.5

Start by clicking the Tools tab and opening the Administrator Settings panel.

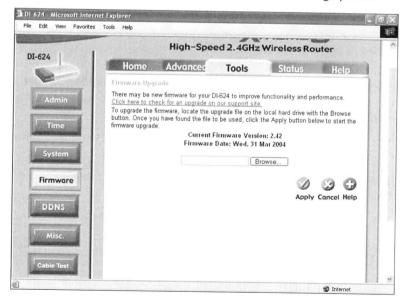

FIGURE 15.6

The Firmware Upgrade panel shows you the current firmware version, providing a link to information about firmware updates and a mechanism for loading a new firmware file.

On the D-Link support site, enter your access point's model number as shown in Figure 15.7. Click the Go button.

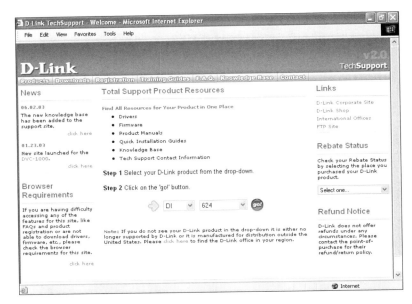

FIGURE 15.7

After you've made note of your current firmware version, you can search the vendor's site for firmware upgrades to download.

Depending on the results of your search, check to see if there is an updated version of the access point's firmware (compare the available versions with the firmware version that you noted on the Firmware Upgrade panel).

If a more recent version of the firmware exists, download it to your computer, noting the location in which the firmware file was saved.

Next, go back to the Firmware Upgrade panel and click the Browse button. Locate and select the updated firmware file that you saved on your computer.

Click Apply. The Firmware update will be loaded in to your access point.

Access Control—Your LAN

Access control facilities that are built in to your access point router combination have a powerful capability to control many aspects of the way your private network, or LAN, operates. The most important of these aspects have to do with the security of your LAN and how computers within your LAN connect to it.

This section explains the mechanics of the most important access control settings, using the D-Link and Linksys access point router combo units as fairly typical examples. Note that overall security concerns (and checklists regarding the actions you should take) that impact a wireless network are discussed in Chapter 18, "Securing Your Wi-Fi Network."

Changing the Administrative Password

In Chapter 13 I advised you not to worry about the administrative password for your wireless access point when first getting your wireless network working. The rationale behind this was it was important to get your network up and running without adding complications. I advised you to leave the password at the default, usually blank or `admin`.

Now that your wireless network is up and running, run — do not walk — to change the administrative password associated with your access point.

Being able to make administrative changes to your access point effectively gives an intruder the ability to manipulate your private network (and even potentially to lock you out of it). So you should change the administrative password. The good news is that to gain access even to an unprotected access point first requires the ability to access your network. The wired part of your network is protected to some degree by being on your premises and under your control. The unwired portion of your network should also have some level of protection if you've implemented encryption as I suggest.

This "perimeter" defense should not make you feel warm and gushy inside. An unprotected access point is still a security risk. It is very easy to change the password, as you can see in Figure 15.8, which shows the D-Link unit's Administrator Settings page. There's very little downside other than the need to remember your administrative password.

For example, to change the administrative password for the D-Link unit, open the Web-based administrative utility. Next, click the Tools tab. Enter your new password for the administrator, confirm it, and click Apply.

As you likely know, you should be careful to choose a password that cannot be guessed too easily. Good passwords contain both letters and numerals, with no recognizable patterns or phrases. For more information about choosing a "good" password, see Chapter 17, "Protecting Your Mobile Wi-Fi Computer."

> **NOTE**
>
> In the case of the D-Link, firmware must be updated using a wired (not a wireless) connection. This is pretty typical.
>
> Be careful with firmware updates: Generally all the settings in the access point are lost, and it is returned to its original, factory default state. So make sure that you've noted all important settings, such as information needed to connect with the Internet, before you perform a firmware update. It's also a good idea to make sure that you archive and keep previous firmware updates until you are sure that the new one is working perfectly.

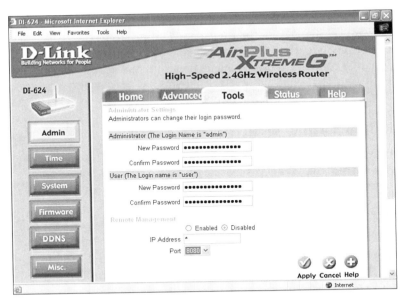

FIGURE 15.8

You should change the administrative password of your access point once your network is up and running.

Understanding Wireless Security

Wireless security primarily consists of two components: authentication and encryption.

As I've previously mentioned, I highly recommend that you protect your wireless network by using *encryption*, which is the process of encoding to prevent an unauthorized user from reading or changing data. This section explains some of the encryption options available for your wireless network and your wireless computer.

Figure 15.9 shows the Wireless Settings panel, used to configure encryption for the D-Link wireless router.

To open the Wireless Settings panel, with the administrative application running, first click the Home tab, and then click the Wireless button on the left side of the screen.

Authentication is the process of ensuring that an individual is who he or she claims to be; note that authentication confers no access rights as such to an individual. (Access rights are handled by the security process known as *authorization*.)

As you can see in Figure 15.9, there are a number of different authentication options for encryption for your wireless access point.

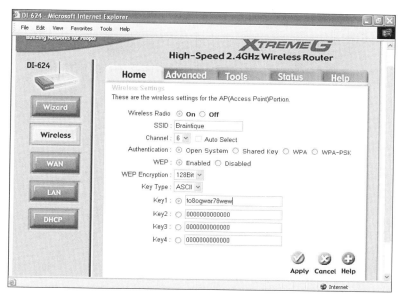

FIGURE 15.9

The Wireless Settings panel in the D-Link administrative utility is used to set encryption.

In Open System authentication, the access points checks to see that a wireless computer connected to it "knows" the encryption key to authenticate it. Encryption keys can be entered as hexadecimal or ASCII strings of text. The wireless computer must enter the key expected by the access point, as shown in Figure 15.10, to be authenticated and to connect.

Shared Key authentication works in much the same way as Open System authentication, except that the authentication process is a bit more complex: The access point sends out random bytes, which the wireless computer requesting access must encrypt using the shared key and send back to the access point. The access point then decrypts using the shared key and verifies that the result matches the original.

The advantage of the shared key scheme is that it proves the wireless computer is using the same key. However, it does give someone with snooping tools more information about your encryption key; so in the context of a wireless network, it is potentially less secure than the simpler Open System authentication.

Shared Key authentication works in the same way as Open System authentication, with a key being entered in the access point (similar to that shown in Figure 15.9 except that Shared Key is selected as the authentication method) and the same key being entered in the Centrino laptop (similar to that shown in Figure 15.10) except that Shared is selected from the Network Authentication drop-down list rather than Open.

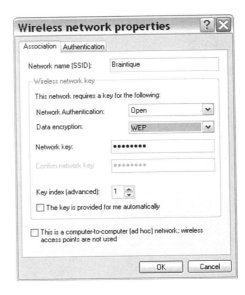

FIGURE 15.10

A wireless computer must enter the key expected by the access point to be authenticated.

WPA, or Wi-Fi Protected Access, authenticates users based on a key that changes automatically at a regular interval using a special server called a Radius server. To use WPA, you need to provide your access point with the location of the Radius server on your network as shown in Figure 15.11.

Of course, you must configure the Radius server used for authentication. Finally, you set the Network Authentication drop-down list to WPA in the Wireless network properties window of your laptop.

If all this sounds like overkill for a small home or office network—particularly buying, configuring, and maintaining a separate server for authenticating wireless clients— you're right; it probably is. So WPA authentication should probably only be considered for situations in which security is important and the resources are available. WPA is probably not the right choice for protecting your Quicken data files, photographs, and Word documents, supposing that you are not a professional network security administrator.

The final authentication choice is WPA-PSK, short for Wi-Fi Protected Access–Pre-Shared Key mode. This is probably the best choice for a home wireless network if all the devices on the network support WPA-PSK.

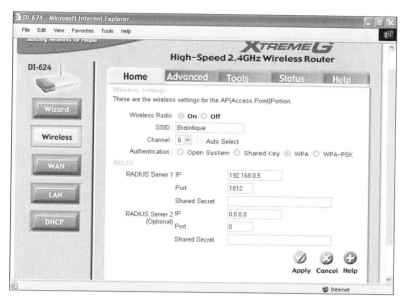

FIGURE 15.11

To use WPA, you must provide the location of the Radius server on your network.

WPA-PSK is sometimes referred to as WPA home version or WPA lite. As opposed to WPA, it does not require a server. When WPA-PSK is selected as the authentication type, you enter a passphrase using your access point's administrative program as shown in Figure 5.12.

In the Wireless network properties window of your Centrino laptop, select WPA-PSK as the Network Authentication type, TKIP as the Data encryption type, and enter the same passphrase, as shown in Figure 15.13.

WPA-PSK uses the preset passphrase for authentication and dynamically generated keys, negotiated between your access point and the wireless computers using it, for encryption. It is probably the best choice for authentication short of full WPA. If you don't want to use WPA-PSK — one reason might be because some older wireless computers do not support it — Open System is the best bet.

If you do choose Open System authentication, make sure that WEP is enabled in your access point (as shown in Figure 15.9). Unless you have older computers using your wireless network that only support 64-Bit encryption, you should select 128-Bit encryption. (It is a little harder to crack.)

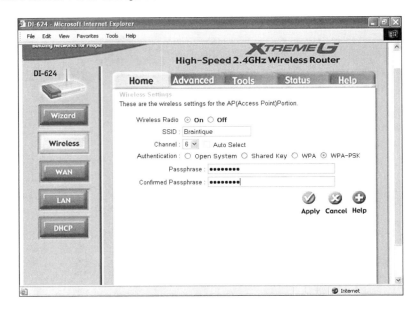

FIGURE 15.12

To use WPA-PSK, you set a passphrase in the access point.

FIGURE 15.13

A wireless computer must enter the passphrase expected by the access point to be authenticated using WPA-PSK.

Finally, multiple keys are used with Open System and Shared Key WEP to vary the encryption key automatically. If you enter multiple encryption keys in your access point, you will need to do so for each wireless computer as well.

Don't Broadcast the SSID

You can choose not to broadcast the network name (or SSID) of your wireless station. By not broadcasting the SSID, your network will not automatically appear in the Available or Preferred networks list in the Wireless Network Connections Properties window of your Centrino laptop. This provides an additional smidgen of security—with SSID broadcasting blocked, it's easy to connect to your wireless network if you know its name, but you must know its name in advance and can't just scan for it.

On the D-Link unit, to turn off SSID broadcasting, with the administrative program open, click the Advanced tab. Next, click the Performance button on the left. The Wireless Performance pane, shown in Figure 15.14, will open. Choose SSID Broadcast: Disabled (shown about midway down Figure 15.14). Click Apply.

> **NOTE** WPA offers the most authentication and encryption protection, but it is too much hassle and expense for anyone but a corporate user. WPA-PSK and Open System WEP are the next best choices, but you should realize that they do not provide complete protection from a motivated and knowledgeable intruder and consider some of the other security measures outlined in Chapter 17.

> **NOTE** The default arrangement is usually to broadcast your SSID to make it easier to connect to a wireless access point.

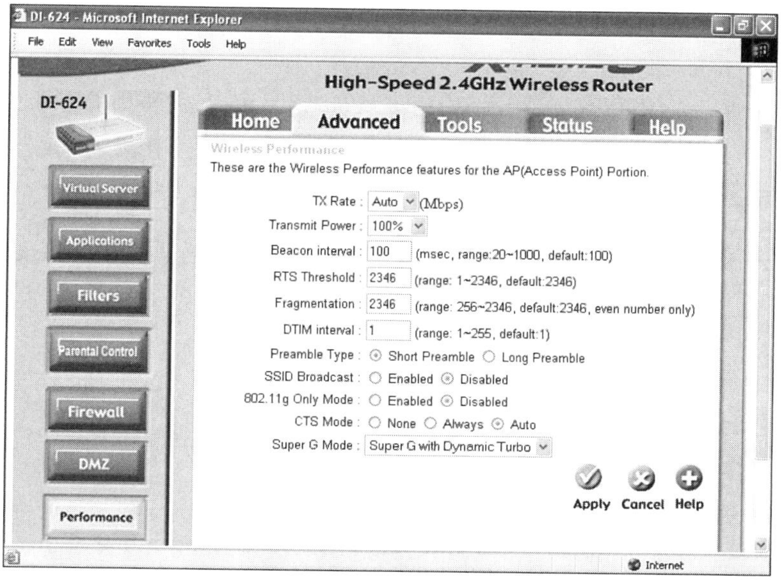

FIGURE 15.14

If you disable SSID broadcasting, users must know your network name to access your wireless network.

TIP
You might need to search your access point's configuration utility to determine how to disable SSID broadcasting for your particular model.

If you open the Wireless Network Connection's Choose a Wireless Network window on your Centrino laptop, as you can see in Figure 15.15, the SSID that was turned off in Figure 15.14 is not listed. (It was named Braintique.)

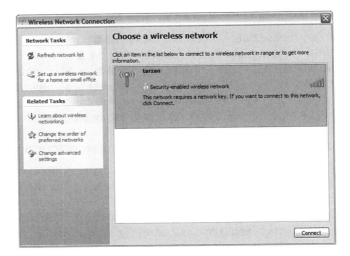

FIGURE 15.15

With SSID broadcasting turned off, you won't see the network name in the Choose a Wireless Network list.

To add a wireless network that doesn't broadcast its SSID, perform the procedure described in "Connecting To the Wi-Fi Network" in Chapter 13. When you get to the Wireless Network Properties window, shown in Figure 15.16, type in the wireless network's SSID and other required information (see "Understanding Wireless Security" earlier in this chapter).

TIP
You should now be able to connect to a new profile containing the station that doesn't broadcast its SSID. You might have to refresh the list of wireless networks (by clicking Refresh Network List on the Choose a Wireless Network pane of the Wireless Network Connection window). If this still doesn't do the trick, try rebooting your laptop to get it to connect to the wireless network whose SSID is not broadcasting.

Remote Management and Providing Virtual Services

Remote management is shown in its usual default state (which is disabled) in Figure 15.17 on the Administrator Settings page of the D-Link's configuration utility.

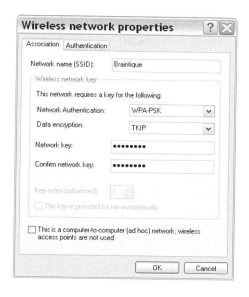

FIGURE 15.16

You can enter the name for a wireless network whose SSID does not show in the Choose a Wireless Network list.

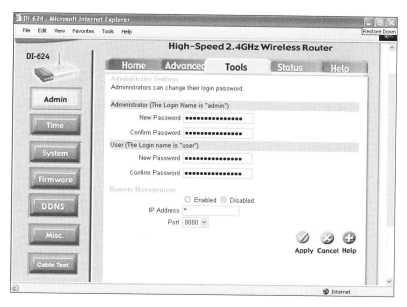

FIGURE 15.17

By default, Remote Management is usually turned off.

> **NOTE**
>
> Some wireless access points provide an additional remote management feature that disables access to the access point administrative utility except from wired computers (and not from wireless computers). This provides a measure of security because wired computers must be physically connected to your network, whereas wireless computers can be operating outside your premises.

Remote management provides the ability to administer your wireless access point remotely, from the Internet. Obviously, this can be useful in some situations. Equally obvious, allowing your private network settings to be changed introduces a security risk.

To remotely manage an access point, it needs to be assigned a fixed IP address (so that it can be accessed). This IP address is entered in the field shown in Figure 15.17. In other words, you are unlikely to be able to use remote management if your ISP provides you with a dynamic IP rather than a fixed one.

Setting up a virtual server is a way to provide users with access to some of the functionality available on your private network without exposing a security hole quite as egregious as allowing Internet administration of your access point.

Figure 15.18 shows the D-Link administrative mechanism for settings up virtual servers.

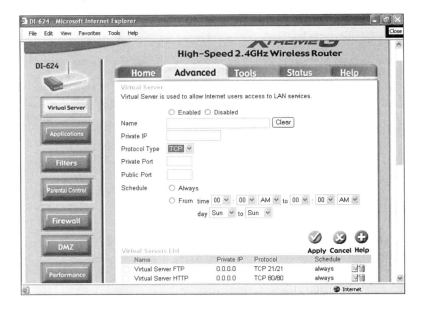

FIGURE 15.18

You can set up virtual servers to allow external users "safe" access to some of the functionality offered by your private network.

You'll find the Virtual Server interface on the Advanced tab of the D-Link administrative utility. This interface allows you to enable various common kinds of services you might want to provide, such as FTP, HTTP, and NetMeeting.

By enabling virtual services, calls directed to the public IP of your access point are redirected to appropriate private IP within your network that can handle the request. For example, FTP requests are redirected to whatever IP you have set up within your private network as an FTP server.

Although providing virtual services necessarily introduces security risks to the assets within your provide network, making these services available in a gated fashion helps contain the risk. In the context of your network and needs, you'll have to decide whether the risk is worth it. On balance, it is unlikely that most small home or office users will want to set up virtual services, but you should know that the possibility exists.

> **TIP**
>
> For a virtual service to be very useful, the public face of your network should be reached via a static (rather than a dynamic) IP. Otherwise, users of the virtual service won't be able to find it!

Using MAC Filtering

A friend of mine who professionally administers wireless networks has told me that his preferred method of securing access to his wireless networks is to deploy MAC filtering.

MAC filtering is a way to enforce security on a Wi-Fi network at a deeper level than any authentication scheme. It's problematic to administer if you expect to often add nodes on your network—for example, when a friend visits with a wireless laptop.

Each device on a Wi-Fi network has a MAC (Media Access Control) address, or unique identification number baked in to the hardware. (See Appendix A, "Wireless Standards," for more information about the MAC layer.) As a matter of fact, every device on your network has a MAC address. In addition to the wireless radio baked in to your Centrino laptop, this goes for your wired network cards, your access point itself, any hubs or switches on your network, USB Wi-Fi devices, and so on.

The idea behind MAC filtering is to tell the Wi-Fi access point that it can only communicate with the devices on your network that are explicitly identified to it by their MAC address. You go into the access point/router's administrative application and say, "Use these MAC addresses and no others!"

> **NOTE**
>
> MAC filtering is a great security tool. But as an administrative matter, it would probably get out of hand if you have more than a handful of devices using the Wi-Fi network or if you added and deleted devices regularly because of the nuisance of adding and deleting all those MAC addresses.

The only trick to this is that you've got to round up the MAC addresses for all the devices you want to be able to connect to your Wi-Fi network.

The good news is that it is usually pretty easy to determine the MAC address for a device.

For one thing, if you buy a Centrino computer or a wireless device, it is likely that the packaging or documentation will tell you the MAC address of your wireless device.

Running Ipconfig, as explained in Chapter 14 and shown in Figure 15.19, will provide a MAC address for the wireless hardware in a Centrino laptop running Windows XP Home edition.

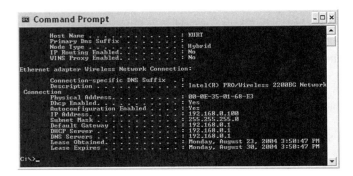

FIGURE 15.19

You can use Ipconfig to determine the MAC address of a network device on your computer.

Windows XP Professional edition also provides a utility, getmac, that shows the MAC addresses of *all* the network devices on your system. (It is a bit dicey knowing which is which if you have more than one.)

To run getmac, choose Command Prompt from the Accessories group in the Start menu to open a command window. With the command window open, type **getmac** at the prompt and press Enter.

As you can see in Figure 15.20, the program will display the MAC addresses of the devices on your system.

You can also use getmac to find the MAC addresses of devices running remotely on your network by supplying the program with the IP or network name of the remote device.

After you've gathered the MAC addresses for the wireless devices that will use your Wi-Fi network, the next step is to enter them in to the Wi-Fi access point/router.

> **NOTE**
> You can use Ipconfig, explained in Chapter 14, to find the MAC address of devices on your current computer. In contrast, getmac gives you a way to find the MACs for all active devices on your network.

Obviously, this will vary depending on the specific piece of equipment.

Using the Linksys Wireless Broadband Router, open the administrative application. Next click the Advanced tab.

In the Filters pane of the Advanced setup, go down to the Private MAC filter item, and click the Edit button. The MAC Access Control Table, shown in Figure 15.21, will open.

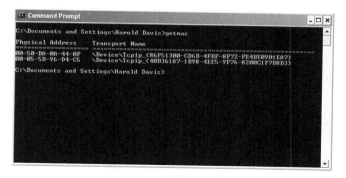

FIGURE 15.20

The getmac *program displays the MAC addresses for the devices on your Windows XP system.*

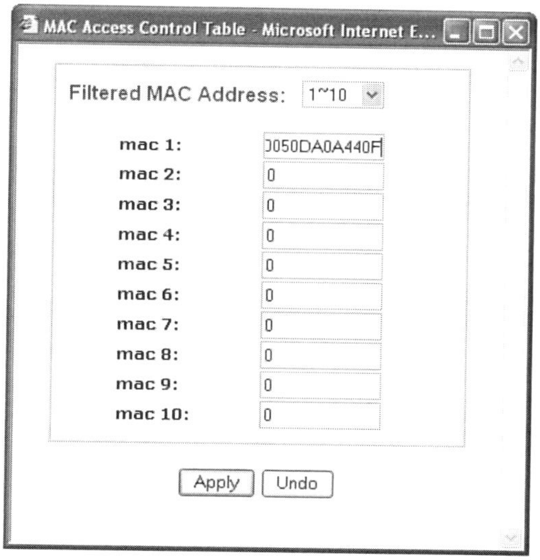

FIGURE 15.21

The MAC Access Control Table is used to enter the MAC addresses of the devices that are allowed access.

Enter the MAC addresses of the devices that are to be allowed access in the table, and click Apply when you are done.

With the D-Link unit, to set MAC filtering, first click the Advanced tab in the D-Link's administrative program. Next, click the Filters button on the left side. Finally, choose MAC Filters (rather than IP Filters) at the top of the pane.

TIP

The MAC Access Control Table shown in Figure 15.21 uses a drop-down list to enter more addresses if you need to add more than 10 MAC addresses.

The MAC Filters interface, shown in Figure 15.22, has considerable power and flexibility.

The D-Link MAC Filters interface can be used to allow or deny access to your private network based on MAC address.

The DHCP Client drop-down list, shown in Figure 15.22, provides a list of the MAC address of all devices currently connected to your access point. You can use this list to add devices to be allowed (or denied) access by the MAC address. Select a device in the DHCP Client drop-down list. Next, click the Clone button. As you can see in Figure 15.22, the MAC address information for the cloned device now appears in the MAC Address fields.

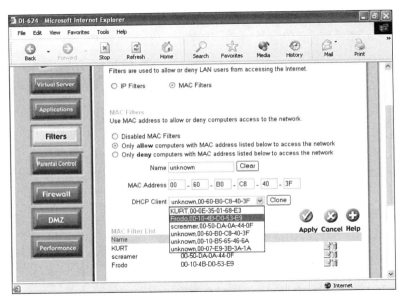

FIGURE 15.22

The MAC Filters interface can be use to allow or deny access based on the MAC address; it can also automatically clone the MAC addresses of devices connected to your access point.

Next, click Apply. The device will be added to the list of devices allowed or denied access using MAC filters (provided that MAC filtering is turned on).

Access Control—The Internet

In the previous section of this chapter, I showed you how to control access to your LAN, or private network. You should want to understand the mechanisms for controlling this access for security reasons. In addition, you might need to open access to your private network in some cases—for example, to make some of its services available over the Internet.

This section explains a different kind of access control—that is, limiting the access of computer on your network to the public Internet.

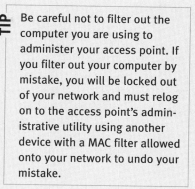

TIP Be careful not to filter out the computer you are using to administer your access point. If you filter out your computer by mistake, you will be locked out of your network and must relog on to the access point's administrative utility using another device with a MAC filter allowed onto your network to undo your mistake.

IP Filtering

IP filtering is used to deny a specific IP, or range of IPs, on your private network the ability to access the Internet.

To open the IP Filtering interface, with the D-Link unit administrative program running, first click the Advanced tab. Next, click the Filters button on the left side. Finally, choose IP Filters (rather than MAC Filters) at the top of the pane. The IP Filters will look like that shown in Figure 15.23. (Of course, it will be different in its layout for different access point administration programs, but the gist is the same.)

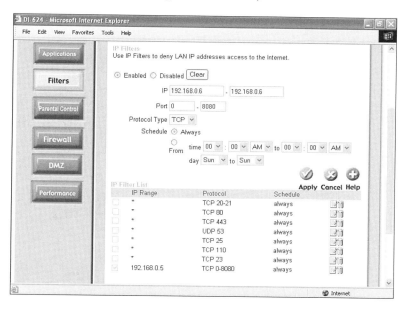

FIGURE 15.23

The IP Filters interface is used to deny computers with specific IPs on your LAN the ability to access the Internet.

To use the IP Filter interface, you enter an IP range (which could just be a single IP), enter a port range (for example, 0–8080), schedule the IP block ("always" is a popular scheduling option), make sure that Enabled is selected, and click Apply. The IP Filter will be added to the IP Filter List at the bottom of the panel.

Here's an example of how you might want to use IP filtering in your home network. Let's say that you've given your 7-year-old son a computer. You want him to be able to access network resources so that, for example, he can use network printers and share files, but you do not want him to be able to surf the Internet. An IP filter fills the bill for this.

You should bear in mind that if you are blocking access by IP, you need to be sure that a device has the same IP each time it boots. This is not how dynamic IP addressing works. So, in the example I just gave, you'd need to open the Internet Protocol (TCP/IP) Properties window for the device connected to the LAN, shown in Figure 15.24.

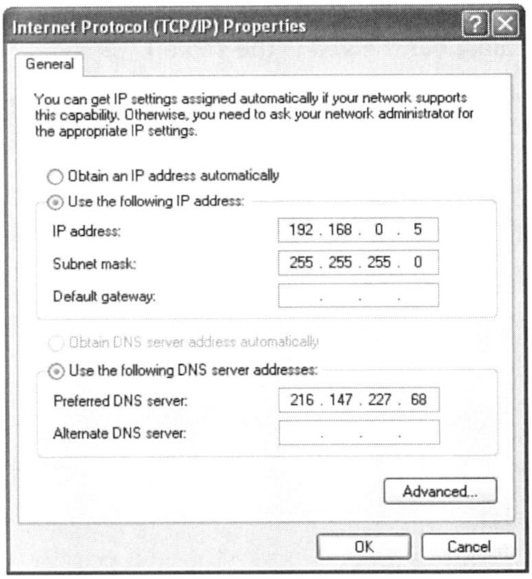

FIGURE 15.24

If you use an IP filter to block a specific computer from accessing the Internet, you need to assign a static IP to that computer.

Make sure that the device is not set to obtain an IP address automatically, and select a static IP for it that is within the range of available IPs for your private network.

Parental Control

Parental Control filters are used to block all users of your private LAN from accessing Internet URLs depending on keywords contained in the URL, or entire domains.

To open the Parental Control interface, with the D-Link unit administrative program running, first click the Advanced tab. Next, click the Parental Control button on the left side.

To block URLs, or Web addresses, make sure that URL Blocking is selected as shown in Figure 15.25.

> **TIP**
>
> Another way to proceed, depending on your access point, is to use static DHCP from the access point to assign a specific IP using the MAC address of the network device that you want to use static IP addressing.

FIGURE 15.25

You can use keywords to block users of your network from accessing Web addresses that contain the keyword in their URLs.

Next, make sure that URL Blocking is enabled. Enter the keyword you want to block. Click Apply.

To block domains, select Domain Blocking as shown in Figure 15.26.

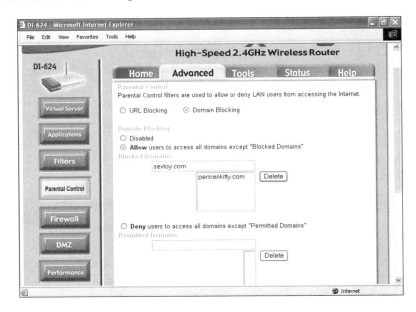

FIGURE 15.26

You can block users of your private net from accessing specific domains on the Internet.

Next, choose to allow users access to all domains except those specifically listed or to deny users access to all domains except those listed. Assuming that you are denying access to specific domains, enter them one by one, clicking Apply after each one is entered. Blocked domains will be added to the list shown in Figure 15.26.

If a user tries to open his Web browser to a blocked domain, or a domain containing a keyword forbidden by URL blocking, he will receive a message similar to that shown in Figure 15.27.

Understanding Dynamic Host Configuration Protocol (DHCP)

You've read a lot about DHCP in the last few chapters, and you might be wondering what it really is. As my grandpa would have said, "Here's the five-cent version."

DHCP (Dynamic Host Configuration Protocol) is an agreed upon standard for assigning variable Internet Protocol (IP) addresses to devices on a network. These variable addresses are called *dynamic* IPs.

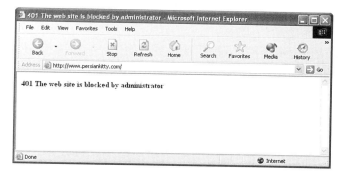

FIGURE 15.27

Users receive a message denying them access if they try to open a blocked URL or domain.

Dynamic addressing makes network administration easier because the software keeps track of IP addresses rather than requiring a human being to perform the task. It greatly simplifies things to be able to add a new computer to a network without the hassle of manually assigning it a unique IP address. For whatever it's worth, you also don't need to have as many IPs with dynamic IP addressing—because not everyone will be on the network simultaneously.

With dynamic addressing, a device can have (and most likely will have) a different IP address every time it connects to the network. In some situations, the device's IP address can even change while it is still connected—for example, when you release and renew your IP settings using Ipconfig.

You might have noticed that so far, I've said "a network," not "the Internet." From a conceptual viewpoint, the Internet is just a great, big, fat network. As a general matter, great, big, fat networks are called WANs (or wide area networks). So from one viewpoint, your router is simply performing a gateway function between your LAN and the WAN that is the Internet. The DHCP servers provided by your ISP give your router a dynamic IP so that it can communicate with the WAN (the Internet). Within your own network (the LAN), the router assigns dynamic IPs to each node on the private network. Using Ipconfig, you can see the IP assigned to each of your computers. A technology called Network Address Translation (NAT) enables a single device, such as your router, to act in this way as an agent between the public network (the Internet) and your local, private network. This enables a single IP address to represent an entire group of computers.

DNS

You might have had to enter settings for a DNS server or set your access point router to obtain a DNS server address automatically. In either case, you are probably curious about what DNS actually is.

DNS is short for Domain Name System (sometimes called Domain Name Service). DNS translates more or less alphabetic domain names into IP addresses. Because domain names are in English (or some other language), they're easier to remember than the tuplets that make up an IP address. For example, it is really easier to remember `http://www.google.com` than it is to remember Google's IP address, `66.102.7.99`, isn't it?

Because the Internet is really based on IP addresses, every time you use a domain name, a DNS server must translate the name into the corresponding IP address. In fact, a whole separate network of DNS servers provide DNS services in aggregate.

The way this network works is that if one DNS server doesn't know how to translate a particular domain name, it asks another one, and so on, until the correct IP address is returned.

DNS is one of the important pieces that makes the Internet function.

More About Mixed Wired-Wireless Networks

In Chapter 13, I explained one good way to configure a mixed Wi-Fi and wired network using a combination Wi-Fi access point/router such as the Linksys Wireless Broadband Router or the Apple Extreme Base Station. This amounted to the simplest possible topology, namely plugging the wired part of the network in to the access point/router as shown in Figure 15.28.

However, the topology shown in Figure 15.28 is not the only way to do things. As I showed you in Chapter 14, complicated networks can get pretty...well...complicated.

One change to the topology shown in Figure 15.28 that sometimes makes a lot of sense is to use a standalone router for your network's routing and DHCP functionality, rather than using the Wi-Fi access point/router to do these things. A reason for this change might be if your standalone router had more sophisticated firewall capabilities than those built in to the Wi-Fi access point/router. (For more about firewalls, see Chapter 18.)

If you made this change, the Wi-Fi access point, whose router functions would no longer be used, would be plugged in to the standalone router along with the wired portion of the network as shown in Figure 15.29.

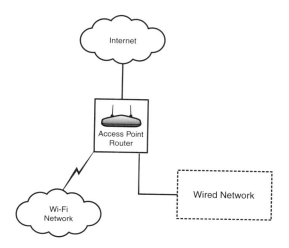

FIGURE 15.28

The easiest way to set up a mixed Wi-Fi and wired network is to plug the wired devices in to the Wi-Fi access point/router.

If you'd like to use your access point/router as shown in Figure 15.29 just as an access point (and not as a router), you need to turn off its DHCP functionality. Figure 15.30 shows disabling the DHCP functionality of the Linksys Wireless Broadband Router using the DHCP tab of the device's administration panel.

Click Apply to keep the change and disable DHCP. If you need to, you can go ahead and set up the SSID and encryption password for your Wi-Fi broadcast.

Next, you can plug the Wi-Fi access point/router in to your standalone router like any other device. Wireless devices such as your Centrino laptop can now happily use the access point, even though the routing capability comes from another router on your network.

> **NOTE**
>
> Quality standalone routers that include firewall capabilities are available from companies including D-Link, Linksys, and Netgear with a wide variety of prices. You can certainly buy one for as little as $30 or $40, but it is doubtful that a router at that price would have any very sophisticated firewall capabilities.

Figure 15.31 shows the interface used to enable and disable the DHCP server that is part of the D-Link unit. Note that in this part of the D-Link's interface, you can also use static DHCP to assign specific IPs to client devices based on their MAC address.

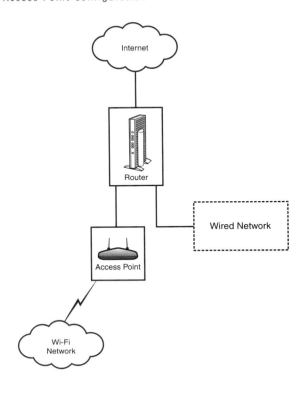

FIGURE 15.29

You might prefer to use a standalone router and plug your access point in to it.

Changing the IP of Your Access Point

> **NOTE**
>
> With DHCP disabled, you might not be able to access the device's administration panel in the normal fashion by entering the appropriate IP in a Web browser. In fact, it might be necessary to use the reset switch on the device (which will re-enable DHCP) before you can access the administration panel again.

As a final thought, you might sometime want to change the IP address of your access point itself on your private network. A reason for this might be because it conflicts with another network device such as a router or another access point.

It's generally quite easy to change the IP address of an access point. Figure 15.32 shows the LAN Settings panel, accessed via the LAN button on the Home tab, used to change its IP.

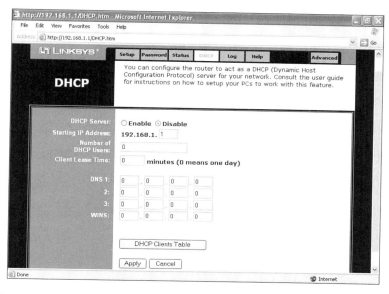

FIGURE 15.30

If you don't want to use an access point/router's routing functionality, you should set its DHCP server to Disable.

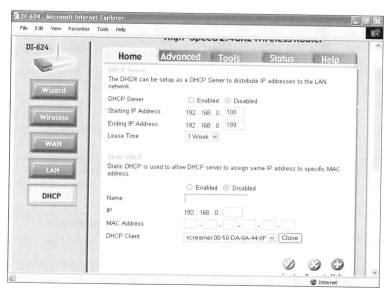

FIGURE 15.31

You can enable (or disable) the DHCP server that is part of the D-Link unit and assign a static IP to a specific MAC address.

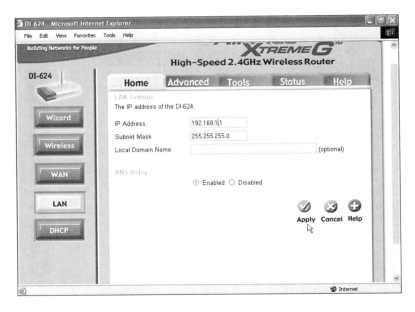

FIGURE 15.32

It's easy to change the IP of an access point on your private LAN.

Simply enter the new IP for the access point, and click Apply.

Summary

Here are the key points to remember from this chapter:

- Dynamic IP addressing enables a variable IP to be used in placed of a fixed IP, making the "bookkeeping" tasks involved in networks, and the Internet, much more manageable.

- A single IP address can be used to represent an entire local area network (LAN) on the outside. Internally, a range of addresses can be generated dynamically for the devices on the network.

- The firmware in your access point can be updated if it becomes dated.

- Authentication is the process of verifying users; encryption means encoding actual transmissions.

> **TIP**
>
> You'll need to use the new IP to open the device's administrative program after it has been changed.

- Unless you can justify a WPA server on your network, you should use WPA-PSK. Or if some of your devices don't support WPA-PSK, use Open System WEP authentication and encryption.

- You can use your access point to expose services and administrative functionality to the Internet.

- You can block specific users from using the Internet and all users from accessing specific sites.

- If you want to use a standalone router, and plug your Wi-Fi access point in to it, you should disable DHCP in the access point (if it provides this functionality).

Adding Wi-Fi Antennas to Your Network

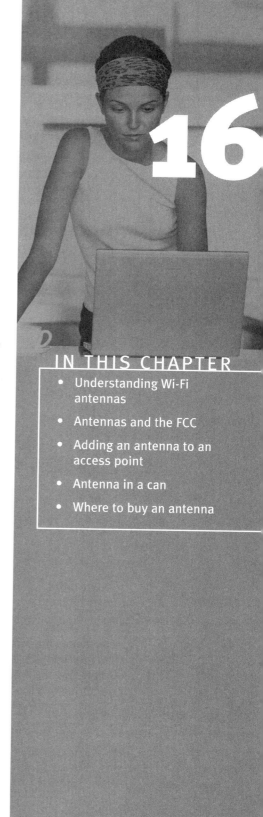

 dding antennas to your Wi-Fi access point can greatly extend the range, coverage, and performance of your Wi-Fi network.

This chapter explains the ins and outs of antennas. What are the best kind to use, and how do they connect? How should an antenna be positioned, and are there other configuration issues? What do antennas cost, and where should you buy yours?

I'll answer these questions and more in this "broadcast blast" of a chapter.

Why Use an Antenna with Wi-Fi?

Depending on your wireless access point, you can add an external antenna to it. For example, in Chapter 12, "Buying a Wi-Fi Access Point or Router," I showed you an example of an external antenna attached to an access point.

Antennas are used with access points to add range to the access point's radio broadcast. They are also an integral part of a Wi-Fi network that covers a substantial area. The use of antennas with your access points can greatly reduce costs in deploying a wireless local area network because you can use fewer access points than you would need without the antennas.

Directional antennas can be deployed with great effectiveness with wireless bridges. For example, you could use wireless bridges combined with directional antennas to connect the wireless access points serving two different buildings in a campus, thus making both buildings part of the same wireless network.

If you are planning to use antennas as part of your Wi-Fi network deployment, it's important to understand the different kinds of antennas that are available. (You also need to make sure to buy access points that come with plugs for external antennas.)

An important part of deploying a Wi-Fi network is doing a site survey, which takes into account the topography and obstacles of the area that needs to be "unwired" and comes up with the best plan and layout for access points and antennas.

Antenna Basics

An antenna is a device that propagates radio frequency (RF) signals through the air. The transmitter (the Wi-Fi card or access point) sends an RF signal to the antenna, which propagates the signal and sends it along through the medium of air.

To some degree, if you want to use your access point with an antenna, you can just go out and buy an antenna (see "Where to Buy Your Wi-Fi Antenna" later in this chapter) and find out how it works. But if you are curious about antennas, here are some things you might want to know about how they work.

When you are thinking about antennas, you need to think about the following characteristics:

- Frequency
- Power
- Radiation pattern
- Gain

Frequency

Antennas used with Wi-Fi need to be tuned to 2.4GHz (802.11b or 802.11g) or 5GHz (802.11a). The frequency of the antenna needs to match the frequency of the radio transmitter.

Power

Antennas are rated to handle a specific amount of power put out by a radio transmitter. In the case of Wi-Fi, this is not a great issue because most antennas are capable of handling the one-watt maximum transmission allowed by the FCC (see "Antennas and the FCC" later in this chapter for more information).

Radiation Pattern

The radiation pattern of an antenna defines the shape of the radio wave that the antenna propagates, or sends into the air. The radiation pattern that all other radiation patterns are compared to is called *isotropic*. In an isotropic radiation pattern, the antenna transmits radio waves in all three dimensions equally so that the pattern represents a ball, or globe, with the antenna at its center. Figure 16.1 is a depiction of the isotropic radiation pattern.

Gain

The amount of gain an antenna provides means how much it increases the power of signals passed to it by the radio transmitter. The amount of gain is measured in decibels (dB), and bears a logarithmic relationship to the power input to the antenna. You should keep in mind that an antenna with a 3dB gain outputs double the power input to it, and an antenna with 6dB gain quadruples the power.

If you look at the specifications provided by antenna manufacturers, you will find gain measured in dBi, or gain in decibels relative to an isotropic radiation pattern. So dBi measures how much "better" a particular antenna is than if using a fictitious antenna with an isotropic radiation pattern, and it is a good measure of how effective an antenna is.

Different Kinds of Antennas

There are many different designs for antennas that Wi-Fi cards and access points can use.

Generally, antennas are either omnidirectional or directional.

Omnidirectional Antennas

Omnidirectional antennas are most commonly used with Wi-Fi cards and access points. These antennas send out radio waves in all directions equally on the horizontal plane but don't send out much in the way of signals vertically. The radiation pattern of an omnidirectional antenna looks like a doughnut, with the antenna in the center of the doughnut, as depicted in Figure 16.1.

Figure 16.2 shows a fairly standard omnidirectional antenna made by Maxrad, Inc., which is intended to work with Wi-Fi and the 2.4GHz spectrum and is mounted externally on a roof.

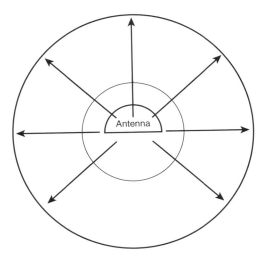

FIGURE 16.1

An omnidirectional antenna sends out signals in all directions on the horizontal plane.

FIGURE 16.2

This omnidirectional antenna from Maxrad is intended for outdoor mounting on a roof.

Directional Antennas

In contrast to omnidirectional antennas, a directional antenna, which is also called a *yagi*, transmits radio signals in a focused beam, such as a flashlight or spotlight. Generally, the manufacturer's specifications give you some idea of the width of the radiation pattern of a yagi. For example, Figure 16.3 shows the radiation pattern for a yagi antenna from Cisco's Aironet division.

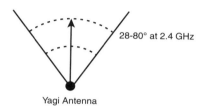

FIGURE 16.3

A yagi transmits RF signals in a focused single direction.

In general, directional antennas have much higher gain than omnidirectional antennas. Furthermore, the higher the gain of the directional antenna, the narrower its beam.

High-gain directional antennas work best to facilitate point-to-point communications—for example, between two wireless bridges, each of which are located on top of a building on a campus. They also can be used to cover a long, but narrow area.

Intelligent use of directional antennas can cut down greatly on the number of access points required to cover an area.

Figure 16.4 shows a yagi (directional antenna) intended for use with 2.4GHz and 5GHz Wi-Fi broadcasts.

Multipolarized Antennas

Antenna development for Wi-Fi (and other wireless technologies) is a hot area right now, with many new developments. For example, a company named WiFi-Plus, Inc., has developed *multipolarized* antennas. According to the company's chief technology officer, Jack Nilsson, these antennas have the capability to propagate and receive signals that are both horizontal and vertical. These models are better than conventional models for going around obstructions. WiFi-Plus's multipolarized antennas can also be used in situations in which Wi-Fi is being broadcast using a directional antenna to a deep valley. A conventional

NOTE In some cases access points come with built-in antennas that cannot be changed. If you buy equipment with a built-in antenna and no way to add an external antenna, you are stuck with what you bought.

directional antenna might broadcast signals that would overshoot the valley, but a multipolarized antenna is capable of broadcasting signals that travel horizontally following the direction of the RF beam, but also can be received down in the valley.

For more information about multipolarized antennas, see the company website, `http://www.wifi-plus.com`.

FIGURE 16.4

This yagi (directional) antenna is intended for use in situations that require high gain.

Antennas and the FCC

In the United States, the Federal Communications Commission (FCC) regulates the use of antennas with Wi-Fi as part of its general regulation of radio frequency devices. You can find the full text of the (rather long and detailed) regulations, which make up Title 47 of Part 15 of the Code of Federal Regulations (CFR) at `http://www.access.gpo.gov/nara/cfr/waisidx_02/47cfr15_02.html`. So that you don't actually have to read this lengthy bureaucratic document, I'll summarize the parts of here it that might matter to you.

These rules create a power limitation for Wi-Fi networks. From the viewpoint of the FCC regulations, the power of a Wi-Fi broadcast is measured in units of equivalent isotropically radiated power (EIRP). EIRP represents the total effective transmitting power of the radio in a Wi-Fi card or access point, including adding gains from an antenna and subtracting losses from an antenna cable. When using an omnidirectional antenna (see "Different Kinds of Antennas" earlier in this chapter for more information about types of antennas) with fewer than 6 decibels (dB) gain, the FCC requires EIRP to be under one watt.

Adding an Antenna to an Access Point

As they say, some guys have all the luck. If your access point does not have a connection for an external antenna, you can't add an external antenna to it. (And you are out of luck!)

On the other hand, if your access point, such as the Apple Extreme Base Station, comes with a connector for an external antenna, it is usually a simple matter to plug an antenna in. (And, you've got all the luck!)

Other access point units fall somewhere in between. For example, the D-Link AirPlus Xtreme Wireless Router that I've discussed earlier in this book, comes with a single, rather small antenna (see Figure 16.5).

FIGURE 16.5

The D-Link unit comes with a single, small antenna.

This antenna can be unscrewed, as shown in Figure 16.6, and a more substantial external antenna used in its place.

FIGURE 16.6

The D-Link's original antenna can easily be unscrewed in order to replace it with a more powerful external antenna.

NOTE Although some Wi-Fi access points intended for home use don't have an external antenna connection, many do. So if you want to have the benefit of extended range from better antennas, shop carefully to find an access point where the antenna is removable.

Adding an external omnidirectional antenna to an access point is essentially a no-brainer. It installs in a jiffy, will improve the range of your access point, and there are no trade-offs. (You don't lose anything except your out-of-pocket cost to buy the antenna.)

Original antennas on access points are pretty small and generally deliver only 2dBi gain at best. Figure 16.7 shows the difference between the original antenna on a D-Link DI-624 and an omnidirectional "rubber duck" antenna with 5.5dBi gain. This larger antenna was purchased mail order from Sharper Concepts for $9.95 plus shipping.

FIGURE 16.7

The original antenna on a D-Link Wireless Router compared to a larger 5.5dBi antenna.

To purchase an antenna that will fit your equipment, you need to know the correct type of antenna connector. Access points with removable antennas almost always have one of the following connectors:

- Reverse Polarity SMA (RP-SMA)—Small, about the size of a pencil eraser, typically found on equipment from D-Link

- Reverse Polarity TNC (RP-TNC)—Slightly larger, about as big around as your pinky finger, typically found on equipment from Linksys

Figure 16.8 shows an example of each connector. If you have difficulty determining which type your equipment uses, consult the website of the access point manufacturer for technical specifications or contact their technical support. HyperLink Technologies has a fairly complete access point-to-antenna connector cross reference list on its website. See `http://www.hyperlinktech.com/web/radio_to_connector_list.php`.

FIGURE 16.8

Examples of an RP-SMA connector (top) and an RP-TNC connector (bottom).

The most gain you will achieve with an antenna directly connected to the access point is about 5dBi–6dBi, as you saw with the rubber duck antenna earlier. To get more gain or to use antennas with special radiation patterns you need a "coaxial pigtail" to convert from the connector on the access point to a more standard antenna connector called Type-N. Figure 16.9 shows a D-Link Wireless Router, an RP-SMA to Type-N pigtail, and an 8dBi omnidirectional antenna suitable for indoor or outdoor use.

Access points and wireless routers are generally not intended for mounting outdoors. That means if you want to put an antenna outside your house, you will need an antenna feed cable to go from the access point inside to the antenna located outside. You need to be sure you get the right cable for the job. There are special types of low-loss coaxial cable for use in 2.4Ghz applications. Suppliers of antennas and antenna pigtails also sell premade cable assemblies with Type-N connectors just for this use. Before purchasing cable, be sure it is specifically intended for use with Wi-Fi equipment. Using the wrong antenna feed cable can result in a significant loss in signal strength, defeating the purpose of special antennas.

Using a coaxial pigtail to convert to the Type-N connector opens the door to a large variety of antennas with special radiation patterns. Selection of more complex, directional antennas is a trickier story than the omnidirectional antennas we have looked at so far. If you are thinking of fitting your access point out with a directional antenna, you need to think carefully about how radio signals will propagate given the terrain you need to cover.

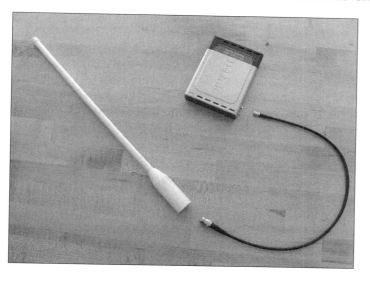

FIGURE 16.9

The D-Link Wireless Router is shown with the black pigtail connector and white omnidirectional antenna.

What the Antenna Should Cost

As you can imagine, there's a great range in price for antennas designed to work with access points. Rubber duck–style omnidirectional indoor antennas are very inexpensive and deliver great value. They can be had for less than $20 from many online suppliers.

Omnidirectional antennas intended for outdoor use start at about $50 and can run you $200 and up.

Directional antennas intended for external use run a little more than omnidirectional antennas. Highly specialized directional antennas can be very expensive indeed.

If you are interested in the multipolarized antennas mentioned earlier in this chapter, you can probably expect to pay a bit more for this feature. A omnidirectional multipolarized antenna intended for external use runs about $150. Directional multipolarized antennas, often used for point-to-point back haul traffic, cost between $200 and $900.

Of course, like everything else related to technology, many antennas can be found for sale on eBay. If you've read this chapter, and understand Wi-Fi antennas, you should be able to buy one safely on eBay and save some cash.

The Antenna in a Can

If springing for a commercial antenna is too expensive for you, you can always consider building a "Cantenna." The total cost of a Cantenna, which uses an old can as its primary part, is less than $5. There's a whole subculture that has formed around building Wi-Fi antennas from cans. Raw materials have ranged from Pringles containers (see Figure 16.10) to (the rather more successful) tin can models. There's even a company, Cantenna, that now sells commercial antennas made from cans.

FIGURE 16.10

This Cantenna made from a Pringles container is a directional yagi-type Wi-Fi antenna based on a design by Andrew Clap.

Besides the tin can, which should be between three and four inches in diameter, the only things you need to build the Cantenna are

- About 1.25 inches of 12-gauge copper wire

- Four nuts and bolts

- A Type-N female connector

> **NOTE**
> As I noted earlier (see "Antennas and the FCC" earlier in this chapter), homemade antennas are probably not compliant with FCC regulations.

The Type-N connector is the standard connector for antennas and will be used to connect it to your Wi-Fi equipment. The cost for this connector is $3–5, which is where I got the original figure of less than $5. You can buy this connector from any of the sources I mentioned as selling radio pigtails (or go to your local Frye's or electronic hobby store such as Radio Shack).

You drill holes in the can for the connector and for the nuts and bolts. Their placement is important because it determines the propagation characteristics of the antenna.

The copper wire is attached to the back of the connector and soldered to the tin can.

The best how-to article I've seen that describes creating a Cantenna, "How to Build a Tin Can Waveguide Antenna," is by Gregory Rehm and can be found at `http://www.turnpoint.net/wireless/cantennahowto.html`. The article includes a calculator to help you determine where to place the connector and nuts and bolts.

You can also purchase a premade Cantenna for $20 (this one meets FCC regulations) from Super Cantenna at `http://www.cantenna.com`. (They sell pigtail connectors besides the Cantennas.)

Where to Buy Your Wi-Fi Antenna

As I mentioned earlier in this chapter, if you want to buy an antenna cheaply, you might have good luck on eBay. I recently tried the terms "RF antenna" and "Wi-Fi antenna" on eBay and found quite a few items for auction, with almost none priced above $25.

If you decide to buy your antenna in a more conventional way, you should go to a company with a Web presence that specializes in RF antennas and related equipment. These companies include

- CushCraft Corporation: `http://www.cushcraft.com`
- Fleeman, Anderson & Bird: `http://www.fab-corp.com`
- HyperLink Technologies: `http://www.hyperlinktech.com` (wholesale only)
- Jefa Tech: `http://www.jefatech.com`
- Maxrad: `http://www.maxrad.com`
- Sharper Concepts: `http://www.sharperconcepts.com` (a retail subsidiary of HyperLink Technologies)
- Til-Tek Antennas: `http://www.tiltek.com`
- WiFi-Plus: `http://www.wifi-plus.com`

Summary

Here are the key points to remember from this chapter:

- Antennas are omnidirectional (all over the place) or directional (a focused beam).
- If your Wi-Fi access point has a connector, you can easily add an external antenna and greatly extend its range.

- Access points with removable antennas use either RP-SMA or RP-TNC connectors. Determine the type used by your equipment before buying an antenna.

- Omnidirectional rubber duck antennas don't cost very much, and they can add a great deal of range to your Wi-Fi network.

- Directional antennas are usually used to broadcast from one point to another point; you need to think carefully about deploying directional antennas with your Wi-Fi network.

- Directional antennas require a pigtail coax cable to convert from the small antenna connector on the access point to a Type-N connector.

SECURING YOUR COMPUTER AND NETWORK

PART V

Protecting Your Mobile Wi-Fi Computer

Security is a big concern for mobile Wi-Fi users, as it ought to be. It's a fact of life, however, that most problems with mobile computing do not have to do with technology, but rather with human interactions. Perhaps someone sees you entering confidential information in your Centrino laptop while you are "Wi-Fiing" in a hotel lobby or a crowded airport waiting area. Or, it can be as simple as theft of a laptop computer. These issues are not much different because your laptop is equipped with Wi-Fi. (Someone can still read confidential information over your shoulder even if you are not unwired.) But it is a fact of life that mobile computers equipped with Wi-Fi do get out and about more—so security is an even bigger concern for them than for the run-of-the-mill non–Wi-Fi laptop. It is also the case that special security risks are involved with using a laptop at a public hotspot.

This chapter explains the dos and the don'ts of traveling with a Wi-Fi–enabled laptop. I'll tell you what you can do to protect yourself and your equipment. I'll also tell you what software you should be running to Wi-Fi with the best of them—for fun and profit, but most of all, safely.

"Social" Engineering

Social engineering is a term for tricking a person in to revealing his password or other confidential information

A classic social engineering trick is to send email claiming to be a system administrator. The email will claim to need your password for some important system administration work and ask you to email it back. Often, the email will appear to be from a real system administrator, and it will be sent to everyone on a network, hoping that at least one or two users will fall for the trick.

You can also be scammed for your password via telephone. In fact, theft of credit card or identity information via "dumpster diving" (or from a restaurant credit card receipt) are examples of social engineering that do not involve technology or the Internet.

Another common trick used by social engineers is sometimes called "shoulder surfing." This is when someone reads your login information, password, or other confidential information over your shoulder.

Wi-Fi users are particularly vulnerable to shoulder surfing. The best defense is to be alert and very careful if you think someone might be looking over your shoulder. If you think someone has read your password, you should change it (or get it changed) immediately. For example, if you think someone might have read your T-Mobile Hotspot password over your shoulder as you entered it in a crowded hotel lobby, you can use the T-Mobile personal preference page to change your password or contact T-Mobile technical support right away by email or telephone.

> **TIP**
> The best passwords are long (at least six characters and digits) and contain both letters and numbers. If a password is very easy to remember, it is probably not that strong a password.

If somebody is watching you when you type in your password, you should move away, or ask him not to look while you log in. It's not polite to read someone else's password, so you shouldn't worry about being impolite yourself when you ask someone not to read it.

Another form of social engineering is guessing your password. You should try to use passwords for logging on to Wi-Fi networks, and passwords in general, that are hard to guess. You should realize that people can find out things about you from public records, such as your date of birth, the names of your children, and so on. So publicly available information about you should not be used for passwords because it can be guessed fairly easily. (For more about passwords, see "Using Password Protection" later in this chapter.)

Social engineering is the biggest threat to computer security, Wi-Fi–enabled and otherwise. The best defense is awareness of the problem and alertness for possible security intrusions.

Physical Lockdown

The physical theft of mobile computers is a pretty big problem, with around 400,000 laptops a year stolen in the United States.

Like other kinds of computer crimes and security breaches, in a great many physical mobile computer thefts, insiders are responsible. Typical insiders include employees, temporary workers, and contractors.

The moral is to be leery about leaving your laptop lying around, either in the office or when you are traveling. This sounds like pretty obvious advice, but what if you just don't want to lug it around with you—for example, to go on a bathroom break during a convention?

A common and relatively inexpensive security device to deal with this kind of situation is the cable lock. The manufacturer of the cable lock provides a way of attaching the lock to the computer. (Often the lock plugs in to a port on the laptop, with a security mechanism preventing its removal without the key.) The cable then loops around a stationary item, such as a desk leg. Except in extreme security situations, it is unusual to chain a laptop to a human being!

Cable locks can be had for as little as $20 to $30. Probably the best known cable lock manufacturer is Kensington, `http://www.kensington.com`. In some cases, the manufacturer of the cable lock guarantees the laptop attached with the cable lock.

The problem with cable locks is that they can easily be cut using bolt cutters available in any hardware store. To add another level of security, you can use a cable lock alarm, such as the Defcon, made by Targus. Targus, `http://www.targus.com`, best known for its mobile computer cases, makes a number of different cable lock alarms for as little as $40. These alarms make a huge racket when the cable is tampered with.

Targus also makes a PC Card, the Targus Defcon Motion Data Protection (MDP) PC Card, that slips into the PC Card slot on your laptop. This card, which sells for about $100, provides double-barreled protection. First, it sounds a loud alarm in response to motion (so it works as a physical theft inhibitor). The card also encrypts the computer, with PIN access. (This encryption inhibits data theft as well as physical theft.)

When the alarm has been triggered (because the card encounters unauthorized motion), a second, 16-digit PIN is required to gain access to the computer's operating system and files.

If you are going to be carrying around important, confidential data on your Wi-Fi–enabled mobile computer, this sounds like a pretty good investment to me!

Quite a few solutions along the lines of the Targus MDP card get more and more complex. Some of these schemes include biometric scanning devices—to authenticate you as the owner of your mobile computer. In other schemes, wireless technology is used to maintain a series of "leases" that keeps the mobile computer going. If the mobile computer fails to obtain a lease for a certain period of time, it stops working, and encryption is engaged. With these schemes, generally a cell phone call can also trigger arming of the defense mechanisms.

Companies that sell sophisticated defense systems along these lines include CoreStreet, Digital Persona, Keyware, RSA Security, and Vasco.

> **NOTE**
> Sophisticated data protection schemes might protect the data on a mobile computer, but they will not prevent the theft of the physical machine. Even if a machine is data locked, the victim of the theft is unlikely to ever get the machine back. So keep a watchful eye on your mobile computer!

Typically, these are complex (and costly) solutions, more suitable for an enterprise than for an individual. But if you are involved in travel with mobile computers that store important and sensitive information, you might want to consider taking this next step.

Using Password Protection

In a mobile computer equipped with Wi-Fi, you can (and should) password protect operating systems such as Windows XP. This makes it a great deal more difficult (although not always impossible) to boot up your computer without knowing the password.

> **NOTE**
>
> For technical information about the security level provided by BIOS passwords, and what is involved in cracking them, see http://www.heise.de/ct/english/98/08/194/. The article also tells how the author used BIOS password protection as a disciplinary measure with his 10-year-old daughter ("We used to give them a slap on the wrist, now we lock them out of their computer!") and how easy it was for his daughter to circumvent the measure.

You can also set a password in the BIOS of most computers. This provides a better level of security than an operating system password, but it is also far from absolute.

To set a BIOS password, you must enter the BIOS screens for your computer. This is done during the boot-up process when you've turned the computer on, generally by pressing a key (such as the Delete key) or key combination while the computer is booting up.

I've already mentioned that you should take care to pick a password that can't be easily guessed by someone who knows a little about you. In addition, proper password management requires some other steps, including

- Choosing passwords that meet certain technical characteristics
- Changing passwords on a regular basis

The long and short of this is that as an individual with a Wi-Fi—equipped mobile computer, you should certainly password protect it before you take it on the road. (Don't forget the password!) But for an organization, real costs are involved in password protection. These include implementing password management procedures, as well as dealing with the inevitable user who loses his password.

> **NOTE**
>
> If you set a BIOS password, make sure that you don't lose it. Despite the various back doors described in the Web reference that I mentioned previously, retrieving a lost BIOS password can be very difficult (and impossible in some cases).

In other words, even a seemingly easy security solution such as password protection has its costs (at least for an organization). So with any security mechanism, it's worth toting up the value of the data that needs to be protected and seeing if the trouble is worth it.

Creating a Password in Windows XP

You can easily password protect your mobile computer. To do so, open the User Accounts applet from the Windows Control panel.

In User Accounts, click your logon name. Next, click Create a Password. The Create a Password for Your Account pane of the User Accounts window, shown in Figure 17.1, will open.

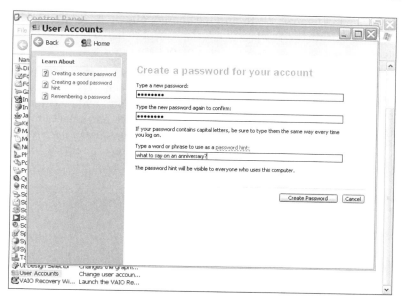

FIGURE 17.1

It's easy to create a password for your user account.

Enter and confirm your password. You can also provide a password hint if you'd like. (This will appear on the Windows Welcome screen when your computer boots up.)

Click Create Password to accept the new password.

File Sharing

We tell our children that sharing is good, but when it comes to computers, running with sharing turned on can pose a security risk.

If you are connecting to a Wi-Fi network with your Centrino laptop—or any network—and sharing is turned on, anyone else on the network can read your files across the network. For that matter, your files can be altered or deleted across the network, as well.

If your folders are private, your files cannot be shared. This might not always be very convenient when you are at home—particularly, as in my case, if you sometimes work collaboratively with a spouse—but it certainly improves security when you connect to a network on the road. In particular, if you are using a public hotspot with your Centrino laptop, it is truly inadvisable to potentially share the contents of your hard drive with the world.

In Windows XP, to turn file sharing off and make your folders private, locate the top-level folder you want to make private. (The folders beneath this one will also be private.)

A good choice would be your entire hard drive, or your folder in the Documents and Settings folder, which contains your My Documents folder (among other things). With the folder you want to make private selected in Windows Explorer, right-click and select Properties to open its Properties window. Click the Sharing tab. On the Sharing tab of the Properties window for the folder, deselect the Share This Folder on the Network check box, as shown in Figure 17.2.

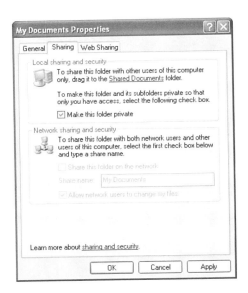

FIGURE 17.2

The Sharing tab of the Properties window for a folder is used to make a folder (and its files) private.

Next, check the Make This Folder Private check box. Click Apply. The folder, folders beneath it within your computer, and the files in those folders are now private and not shared.

If you want to reverse the process when you get home, clear the Make This Folder Private check box, and check Share This Folder on the Network. Your files and folders are now public, meaning shared. If you really want to live dangerously, you can also check the Allow Network Users to Change My Files box, also shown in Figure 17.2.

Using a Virtual Private Network (VPN)

You've logged on to a public hotspot at some great location. This time, let's say that you're connected poolside at some great hotel in a warm location. You can hear the gentle ocean surf not far away, and smell the finger food at the open-air bar. So far, all is well and good.

When you connected to the Wi-Fi hotspot provider, say T-Mobile Hotspot or Wayport, you were probably authenticated. This means that you had to provide a login identification and a password—in part so that the Wi-Fi provider would know who to bill for your time online.

But beyond this authentication, there is no security at a public Wi-Fi hotspot.

If the Wi-Fi service is free and/or put together as a one-off by the establishment you are visiting, there might not even be this level of authentication.

Furthermore, your Wi-Fi transmissions are probably not encrypted. Wireless networks are inherently less secure than wire line networks because anyone can pick up the signals. Without encryption, tapping into your Wi-Fi packets as they are transmitted is like reading plain text. It is not very hard to do from a technical viewpoint.

I don't want you to get the impression that the world is full of people who live to lurk and to pick up your Wi-Fi transmissions. And, after all, every time you hand a credit card to a server at a restaurant there is a security risk—not unlike the security risk with open Wi-Fi.

But you should regard a public Wi-Fi hotspot as fundamentally insecure. If you need to work remotely—as I sometimes do on consulting projects—with information on your home (or, more likely, office) network, this can be problematic.

One solution that can be used to at least make a connection more secure from a public hotspot to your home network is to set up a Virtual Private Network (VPN) by installing a remote access server on your home network. VPNs use a dedicated server to "tunnel" through the Internet and provide a way to communicate securely with your home network, as shown in Figure 17.3.

NOTE

Under Windows XP, you can only change the permissions on files and folders to make them private if the NTSF file system is being used. The NTSF file system is superior to the alternatives, FAT and FAT32 (which are used in Windows 9x/Me and optional in Windows XP), anyhow, so there is little reason not to be running it. If you can't make folders private, your computer is probably not running NTSF. For information about how to convert to NTSF, open Windows Help and Support, and search for NTSF.

You should also know that the details of the process for setting sharing will differ from what I described if you are in a network with a domain controller, rather than in a workgroup network. For details of resetting permission in a network with a domain controller, see "Sharing and Security" in Windows Help.

TIP

For an explanation of authentication and encryption, see Chapter 15, "Advanced Access Point Configuration."

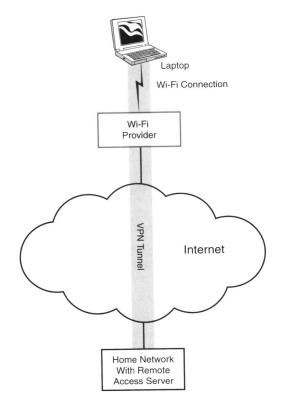

Laptop

Wi-Fi Connection

Wi-Fi Provider

VPN Tunnel

Internet

Home Network With Remote Access Server

FIGURE 17.3

A VPN tunnel creates a secure communications channel across the insecure Internet.

> **NOTE**
>
> As specific examples of security risks that might concern you, this means that someone with the know-how and right equipment could easily sniff passwords that you provide to Web servers, read your unencrypted email, note which websites you visit, and track your online banking and credit card transactions.

This means that using a VPN is a very helpful way to add security to a Wi-Fi connection. After you've logged in to the VPN, you can use the resources of your home network without feeling that security is compromised.

There are a great many vendors of VPN server products, which are mostly geared at the enterprise. For a good source of information about VPNs, and to find the companies that are involved in making the server software, you might want to have a look at the website for the Virtual Private Network Consortium, better known as VPNC, `http://www.vpnc.org`. The VPNC is the trade association for companies that make VPNs.

It's good news that the client software for use with a VPN is baked in to the Windows XP operating system.

To make a connection to a VPN from Windows XP, open the Network Connections applet by double-clicking Network Connections in Control Panel. In Network Connections, click Create a New Connection. (You can find this on the left under Network Tasks.) When the New Connection Wizard opens, click Next.

Choose Connect to the Network at My Workplace, as shown in Figure 17.4, and click Next.

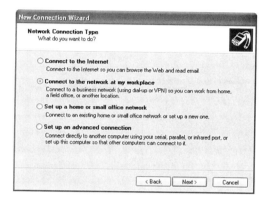

FIGURE 17.4

Choose Connect to the Network at My Workplace to start the wizard that creates the VPN client on your remote system.

Choose Virtual Private Network Connection as shown in Figure 17.5, and click Next.

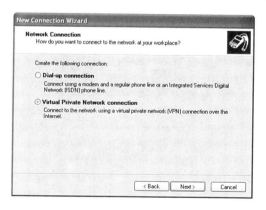

FIGURE 17.5

To create a VPN, choose Virtual Private Network Connection.

You will be asked to provide a name for the VPN, such as myVPN, and the host name, such as bearhome.com, or the IP address of the VPN server.

Click Finish to close the wizard. The VPN will now appear in your Network Connections window, as you can see in Figure 17.6.

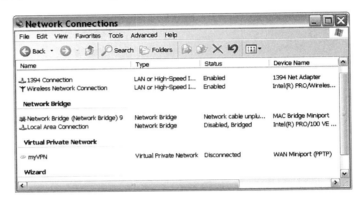

FIGURE 17.6

The new VPN is shown in the Network Connections window.

When you attempt to connect to the VPN, you will be prompted for your VPN username and password so that the VPN remote access server can authenticate you, as shown in Figure 17.7.

FIGURE 17.7

When you connect to the VPN, you will be asked to supply a logon and password so that the VPN's remote access server can authenticate you.

Personal Software for Protection

Any computer that connects to the Internet, or that connects to a network that connects to the Internet, should run some programs for general protection.

Antivirus Programs

The most important category of program for protection is antivirus. Antivirus programs stop viruses from attacking your system and help you recover if you are attacked by a virus.

The leading antivirus products are VirusScan from McAfee, `http://www.mcafee.com`, and Norton AntiVirus from Symantec, `http://www.symantec.com`.

The two important things you must do with your antivirus software (or else it doesn't fully protect you) are as follows:

- Keep your virus definitions up-to-date
- Scan your computer for viruses regularly

The good news is that you can set most modern antivirus programs up to perform these functions automatically—so that you don't have to remember to do them.

Automatically Updating Virus Definitions

For example, with Symantec's Norton AntiVirus program, keeping program code and virus definitions up-to-date is the job of a portion of Norton AntiVirus called LiveUpdate. To set LiveUpdate to automatically check for new virus definitions, click Options in Norton AntiVirus. Next, Choose LiveUpdate. The Automatic LiveUpdate window will open.

In the Automatic Live Update window, shown in Figure 17.8, check the Enable Automatic LiveUpdate box. Next, choose the Apply Updates Without Interrupting Me option. Click OK. Your virus definitions will now be updated automatically.

Automatically Scanning Your Computer for Viruses

To set up automatic scheduled scanning of your computer for viruses, make sure that the main Norton AntiVirus window is open. Next, click Scan for Viruses to open the Scan for Viruses pane (shown in Figure 17.9).

NOTE

As you probably realize, there are some costs (in time and trouble if nothing else) in maintaining a VPN server. This is probably not something that is worth doing unless you have a number of remote users who need to access your home network securely on a regular basis.

TIP

In addition to the antivirus and personal firewall software mentioned in this section, you should consider using software to help protect you from spyware, adware, and other forms of malicious intrusion. See Chapter 4, "Software That Makes the Most of Mobile Computing" for more information.

NOTE

Because new viruses come on the scene so quickly, it is vitally important to keep your virus definitions up-to-date. If your antivirus program comes with an automatic update feature, you should make sure that this is turned on and operational. Otherwise, make sure to go online and update your virus definitions frequently. An antivirus program with out-of-date definitions is almost as bad as no protection at all.

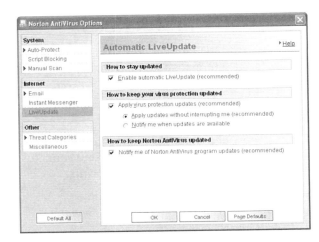

FIGURE 17.8

You can automatically keep your virus definitions up-to-date.

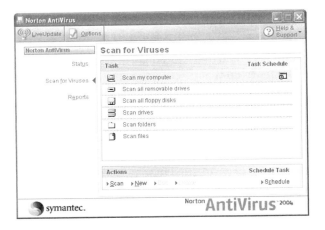

FIGURE 17.9

You can automatically scan your computer for viruses.

In the Scan for Viruses pane, you can click on a task—such as Scan my computer or Scan my drives—to perform an antivirus scan. Sometimes, if you have reason to think you might have been infected, it's a good idea to do this. However, to implement overall protection, you should click in the Task Schedule column to the right of the task you want to schedule.

For example, to engage automatic antivirus scanning of your computer, click in the Task Schedule column to the right of Scan My Computer.

Clicking in the Task Schedule column opens the Schedule tab, shown in Figure 17.10.

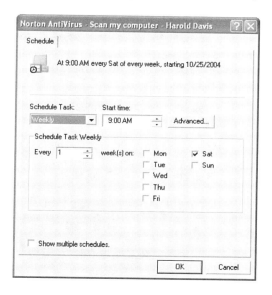

FIGURE 17.10

The Schedule tab is used to engage automatic antivirus scanning.

Within the Schedule tab, select a period for automatic scanning, such as once a week, and a time for scanning during the period (such as Saturday morning, shown in Figure 17.10).

Click OK. It's that easy—you are now automatically protected with Norton AntiVirus, which updates your virus definition files without your intervention and scans your full computer once a week. By the way, if your computer is off when it is set to do a scan, don't worry—the next time you turn it on, you'll be told about the missed event and asked whether you want to start the scan.

> **TIP**
>
> It's best to schedule scans for a time when you're not likely to be making heavy use of the computer because antivirus scanning of your whole computer does take a while and will slow the computer down.

Personal Firewall Programs

A firewall is another type of program useful for protecting your PC. A firewall program protects your resources by filtering network packets. If your computer is connected to a network that is not protected with a firewall, or if you are taking a computer on the road, you probably should use a personal firewall program for added protection.

For the most part, you don't need to do anything to be protected with a personal firewall program other than turn it on. So the best thing about this kind of protection is that it isn't a nuisance to use.

A number of good personal firewall products are on the market, including offerings from McAfee and Symantec that come bundled with their antivirus products. Other personal firewalls worth considering include Personal Firewall from Tiny Software, http://www.tinysoftware.com, Sygate Personal Firewall from Sygate Technologies, http://www.sygate.com, and Zone Alarm, from Zone Labs, http://www.zonelabs.com.

In addition, Windows XP comes with personal firewall software, with the grand virtues that it is free and (if you have Windows XP) you already have it.

To activate the Windows XP personal firewall, just go the Control Panel and double-click the item that says Windows Firewall. To turn on Windows Firewall click the On button as shown in Figure 17.11, and then click OK. That's all there is to it.

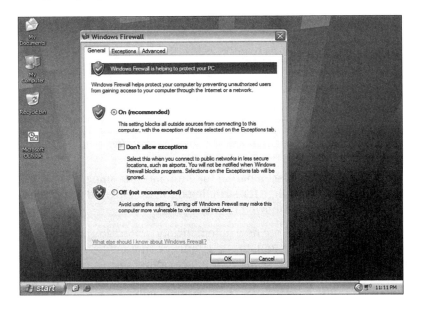

Figure 17.11

It's easy to turn on the personal firewall that ships with Windows XP.

The status of each network connection will now include "firewalled" in the Network Connections window.

Summary

I hope this chapter has helped you learn how to practice safe computing. Using your wireless laptop is no more perilous than the other aspects of life itself if you follow the suggestions in this chapter.

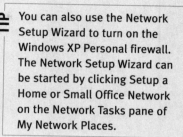

TIP

You can also use the Network Setup Wizard to turn on the Windows XP Personal firewall. The Network Setup Wizard can be started by clicking Setup a Home or Small Office Network on the Network Tasks pane of My Network Places.

Here are the key points to remember from this chapter:

- Protection from security problems is a mindset that requires awareness and alertness.

- Social engineering is the biggest security threat.

- Physical theft of laptops is a big problem, but some devices can help with locking down your equipment.

- You can enhance your safety by password protecting your computer and disabling sharing.

- If you need to access your home network from a public hotspot, setting up a VPN can enhance security.

- You should run antivirus and personal firewall protection on mobile computers equipped with Wi-Fi.

Securing Your Wi-Fi Network

We live in a world that can seem increasingly dangerous both to human beings and their computers. Wi-Fi networks are very convenient, wonderful to use, and easy to deploy (as I explained in Part IV, "Your Own Wireless Network"). But running a wireless network does pose a substantial security risk.

This chapter will help you assess the risk that your wireless network poses and create a game plan for dealing with these risks. I'll also show you how to work with some of the tools that can be used to minimize risks, including firewall and VPN servers.

Understanding the Threat

Make no mistake, the threat is real. If you compare a wireless network with a conventional wired network, the security risks posed by the two are essentially the same with one big additional risk for the wireless network.

The big additional risk is that a wireless network provides no physical security. Essentially, anyone can tap in to a wireless network. In comparison, to hack a wired network, you need a physical connection to the network's wiring, generally meaning that you must have access to the premises, implying an inside job of some sort.

Attacks from the Internet are, of course, a threat to both wired and wireless networks. But in stark contrast to wireless networks, no one can attack a wired network without gaining admittance to your premises. Wireless networks are vulnerable to attacks from people who are not on your physical premises. This means that protection cannot be obtained by physical security measures, but only by implementing appropriate internal management and security measures. A lock on your door should inhibit someone who would like to access your wired network, but it is meaningless to the security of your wireless network.

Another facet of the problem is that the default setup for a wireless access point/router just gets your Wi-Fi network up and running. It doesn't step you through the process of adding any security features, such as encryption, to your network. (See Chapter 15 for this information.)

An astounding percentage of private Wi-Fi networks—some estimates are as high as 80%—are run without any security features turned on. When I scan the immediate neighborhood of my house, I find about a dozen wireless networks (other than my own). Most of these are open, meaning that they require neither authentication nor encryption to use. Judging by the names of the wireless networks ("Linksys," "Netgear," and so on), my neighbors who set them up just used the factory default settings (and haven't read this book). This is a huge security hole.

It's also worth noting that public hotspots typically don't feature any security besides basic user authentication—if even that—because the people running the hotspot want to make it as easy as possible for people to log on.

I don't want to exaggerate the problem. You might quite rightly feel that you have no secrets and that you don't care about giving away access to your files to strangers.

There's some merit to this position. It's likely that no one would really care about most of my files (or your files). In any case, it's probably worth a lot less effort to guard, say, driving directions to your favorite restaurant than, say, the firing sequence for a nuclear warhead.

Every security management issue comes down to a balancing act: Is that which is being secured worth the cost (in time, trouble, and money) of more stringent security? But everyone has something worth safeguarding. For example, you probably really don't want to hand out your Quicken or Microsoft Money data files to strangers. Personally, I don't even want to share pictures of my kids—except those I post on my website myself—with strangers.

The most stringent security of all would ban wireless networking and indeed networking altogether—because whenever there is communication in and out, there is a potential risk. As with human social interactions, every interaction between computers is a risk. That's why truly locked down security measures include removing all physical—wired and unwired—network connectivity (as well as all access to removable media).

But, for most people, taking that kind of step would be not worth the cost. It would be such a nuisance to try to work without connectivity that the security is just not worth it. In a similar fashion, Howard Hughes was probably right that shaking hands with people can spread illnesses, but the extreme measures he took to cut himself off from peo-ple—rubber gloves, becoming a hermit—were not worth the putative benefit of free-dom from the flu.

To more fully be able to perform the security balancing act, I'd like to step back for a minute and look at just what the security threat to your Wi-Fi network is.

If your Wi-Fi network is open, or completely unsecured, someone (whom I'll call the "nefarious evildoer") within broadcast range of your access point, but probably outside your physical perimeter, can become a node on your network. This is sometimes called *penetration* or *intrusion*.

As a node (or client) on your network, the nefarious evildoer can access files on your network.

Access to the file systems on your computers means more than that the nefarious evildoer can read the files. The nefarious evil-doer can also alter and delete them. If the nefarious evildoer is really malicious, your entire system could be wiped out.

> **TIP**
>
> See Chapter 17, "Protecting Your Mobile Wi-Fi Computer," for information about turning file sharing off so that access-ing files, even with network access, is harder to accomplish.

The nefarious evildoer, depending on how you have things set, can also change your network administrative settings. You could get locked out of your own network!

If you haven't changed the password in your access point, the nefarious evildoer could open its administrative panel, assuming (as most access points do) that it uses Web-based administration. The settings could then be changed to defeat whatever security measures are in place.

Of course, most penetration is relatively innocent and is done to obtain Internet access. Yes, the nefarious evildoer just might not have Internet access and want to piggyback (without paying) on yours.

Before you throw up your hands and say, "I don't care. I'm happy to share my Internet connection: it's not going to cost me any more. Besides, sharing is in the spirit of open source, Wi-Fi, and all those good things," you should think about a couple of ramifications.

By sharing your Internet access in this way, you are probably in violation of your agree-ment with your ISP. Okay, so I don't care much about this technicality either. But if some real nefarious evildoer does use your ISP account to launch a Web attack—using a virus or a denial of service campaign—you could be held responsible. At the very least, it could lead to the ISP shutting down your account. Also, if others are using your Internet connection, there's no doubt your connection speed will slow. I don't know about you, but even broadband isn't fast enough for me. I don't want freeloaders gum-ming up the works even more.

> ## FILE-SHARING RISKS
>
> Concerns about losing bandwidth are particularly valid in the case of file sharers.
>
> Another concern in this respect is that file sharers are almost certainly trading in copyrighted information (songs), and the person who is the owner of the connection to the Internet is the one who the RIAA (Recording Industry Association of America) is going to track down.
>
> Child pornography is a lesser concern simply because fewer people traffic in that, but it's still something to think about.
>
> In other words, if you leave your network open, you might be liable (both civilly and criminally) for the actions of freeloaders who use it, as well as the somewhat lesser issue of suffering from diminished bandwidth.

Before you say it's okay with you to have others use your Internet connection because it doesn't cost you anything more, think about whether you would leave the front door to your house open with a note saying, "Come in; use the phone: Local and long distance minutes are free!"

What Steps Should You Take?

The steps you should take depend on how important the security of your personal network is to you. Some people will feel it more important than others to implement comprehensive security measures. But some of the basic security measures you can take are easy and involve little (or no) trouble to set up and very little extra trouble on the part of network users. So everyone should take at least some security measures:

NOTE The measures described in this section cover network security. Besides the measures explained in this chapter, you should also take steps to protect individual computers such as installing antivirus software and personal firewall software, as I explained in Chapter 17.

- Change your network name (SSID) so that it is not the default

- Set your SSID not to broadcast

- Implement WPA-PSK security (preferably) or at least WEP

- Make sure that all the computers on your network are running up-to-date antivirus software

- Change the default administrative and user passwords in your access point

I'll explain these steps in a little more detail in a moment, as well as what to do if your situation calls for greater security than the minimal measures provide. In other words, one key step is to develop a security game plan in the context of your own requirements for security because all serious security measures involve costs and trade-offs.

To come up with a game plan for implementing security on your Wi-Fi network, you should sit back for a moment and see which of these security levels makes the most sense to you.

No-Brainer Security

The measures described in this section are the ones that everyone with a Wi-Fi network should take. (This means you.)

Even if you truly believe that you have nothing of any value on your network, these measures are so easy to implement and no trouble to use, so why not put them in place?

No-brainer security measures include the following:

- Change the SSID, or network name, from its default.

- Set the SSID not to broadcast. If your SSID is not broadcast, it will be harder for a nefarious evildoer to log on (or even know that your Wi-Fi network is there).

- Implement Open System WEP (wired equivalent privacy) or WPA-PSK encryption and authentication if your devices support it.

- Make sure that all the computers on your network are running antivirus software and that the virus definitions are updated weekly. This has more to do with general network protection (and common sense) than it does with Wi-Fi network security, but it is still very important.

- Change the default password for the administrative application for your access point.

> **NOTE**
>
> The National Institute of Standards and Technology NIST), which is a branch of the U.S. Department of Commerce, has prepared a comprehensive white paper about wireless network security. The white paper contains a very helpful and comprehensive wireless LAN security checklist containing 45 items. The items are characterized as "Best practice" (meaning it should be done) and "May consider" (meaning for those who are a little more concerned about security).
>
> You can view this NIST white paper at
> `http://www.csrc.nist.gov/publications/nistpubs/800-48/NIST_SP_800-48.pdf`.

PICK YOUR PASSWORDS WISELY

For security reasons, you should be careful about the passwords you pick. Never use as a password the name of your company, anyone in your family, or your pet hamster. (These are all easy to guess.) For the same reason, don't use your date of birth or street address.

The worst kind of password is one based on personal information that can be easily determined (or guessed). But you should also avoid passwords based on numerical sequences (such as 12345) or words (such as "password"). Better passwords are at least eight characters long and include both letters and numbers: for example, d31TY9jq.

Unless no one ever visits your home or office, don't jot down your important network passwords on a post-it, and then stick the post-it on your computer monitor. See Chapter 17 for more information about choosing a password.

NOTE

Neither Open System WEP nor WPA-PSK encryption should be regarded as the final solution to security problems. With the right tools, both forms of encryption can be cracked in less time than you might think. WPA-PSK is the stronger of the two, particularly if you use an automatically generated 64-bit hexadecimal password string rather than the passphrase mechanism built in to many WPA-PSK implementations (if your equipment allows you to do this). If you choose a short passphrase with WPA-PSK, it might even be less secure than WEP. The bottom line is don't count on either of these forms of encryption as an absolute defense against a determined and knowledgeable attack.

Middling Security

You should consider using some of the middling security measures described in this section if you have some security concerns and are prepared to take some trouble over security, but still don't want to go overboard.

In other words, these measures should make sense to you if your posture is, "I'm reasonably concerned about security, but I don't want to waste too much time on it. I'm willing to go to a little bit of trouble to make my network more secure. I'll do what I should do, so long as it is not too much trouble."

Middling security measures include the following:

- Plan to change your Open System WEP or WPA-PSK encryption key regularly, perhaps once every week.

- Engage MAC filtering. For relatively small networks, MAC filtering is an absolutely excellent, and a relatively painless, way to enforce good security, but it is problematic to administer if you expect drop-in ad hoc wireless users.

- Use the DHCP settings in your router (or access point/router combination) to limit the number of IPs that can be used in your network to the actual number of devices that you simultaneously connected to the network.

Red-Alert Security

Red-alert security measures are intended for use with networks that truly have confidential and proprietary information to protect and are willing to go to considerable trouble and expense. (These networks are perhaps the home of proprietary and confidential information belonging to clients.)

> **NOTE** After you've changed the password on the access point, there might be no way to get it back short of doing a hard reset on the access point (which means that all your settings will get lost). So keep the password in a safe place, and don't lose it.

You should realize that protection at this level is not a one-shot affair: You have to constantly be on the lookout for new vulnerabilities. You'll need to keep surveying your wireless site, keep changing your passwords, and generally just keep on your toes. Expect red-alert security measures to take time and money, to be trouble to maintain properly, and possibly to slow down your network!

If you run your own business, as I do, when people ask about your job description, you might well say "Chief Bottle Washer." If you intent to deploy red-alert security on a wireless network, to that job title you could well add "Network Administrator" and "Wireless Security Expert."

If you really need a high level of security, you should consider not using a wireless network at all or, at the very least, bringing in a qualified wireless network security expert.

> **NOTE** No wireless network can ever be completely secure. Keep any truly confidential information off a wireless network.

Red-alert security measures include the following:

- If your access point allows this, lower your broadcast strength. The lower your broadcast strength is, the less likely a nefarious evildoer outside your network is to be able to intercept it (because it doesn't broadcast outside your premises). The ideal scenario here requires fine-tuning your Wi-Fi broadcast so that it is strong inside your premises but falls off rapidly outside. This can often be accomplished by turning down the transmission power, as shown in the D-Link access point unit in Figure 18.1, combined with clever network design and yagi-type antennas.

- Understand the range of your Wi-Fi broadcasts, and see if there are any obvious vulnerabilities. (A parking lot? A neighbor who hates you?) Performing a physical survey will not only help you understand vulnerabilities, but it will also help you create a network topography and transmission plan that bypasses the problem areas you have found.

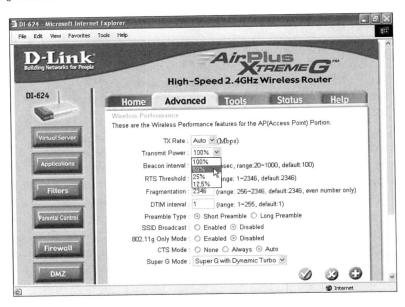

FIGURE 18.1

The transmission power of the D-Link AirPlus access point is cut down to 50% using the Transmit Power drop-down list on the Wireless Performance pane of the Advanced tab.

- Regularly review the DHCP logs provided by your router to see if there are any unauthorized connections. A portion of the log for the D-Link access point is shown in Figure 18.2.

- Turn off wireless access to the access point's administrative application. (This is usually only available with enterprise-class Wi-Fi access points.)

- Use a dynamic, per-session WEP encryption scheme. This requires additional hardware, namely an authentication (or Radius) server.

- If you can't install a dedicated authentication server, authenticate Wi-Fi connections with usernames and passwords using a network directory server (which can be a Windows domain server and need not be a separate piece of hardware).

- Encourage access to your Wi-Fi network via a VPN.

- Create a network topology that uses a DMZ with its own set of firewalls for the Wi-Fi access point. This will isolate the access points from possible attacks. You can beef this up even further by making sure that the access point and the nodes on your wireless network can only communicate via a VPN. In the small office context, a good piece of equipment to use to implement this is the Watchguard SoHo Firebox, which combines a firewall and a VPN and costs about $300.

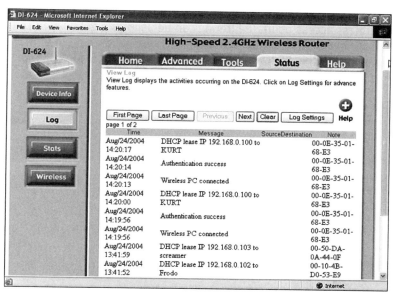

FIGURE 18.2

Regularly checking the DHCP log maintained by your access point for illicit connections is part of red-alert security.

Understanding Firewalls

A firewall is a program that protects your resources by filtering network packets. Firewalls can be run as part of another piece of software. For example, your Wi-Fi access point/router almost certainly provides some kind of firewall capabilities. Firewalls can also be run as individual programs on computers. Finally, sophisticated firewalls can be run on servers dedicated to that purpose, although this generally only happens in enterprise-class setups.

Firewalls enable a network administrator to determine which clients inside a network can access network resources and which *ports* can be used from outside the network to access the network. In case you are wondering, a network port is a logical endpoint on the network. The port number identifies the kind of traffic that uses the port. For example, port 80 is used to connect to a Web server using the HTTP protocol.

Effectively, firewalls can be used to isolate portions of a network topology from the rest of the network and from the Internet. This is another way of saying that you can use a firewall to limit access both to and from the Internet.

Within a wireless access point, IP filtering can be used as a gateway to control access to your private network. In addition, the firewall application built in to many access points (the D-Link AirPlus Firewall Rules screen is shown in Figure 18.3) can be used to let traffic (or not let traffic) pass through the access point.

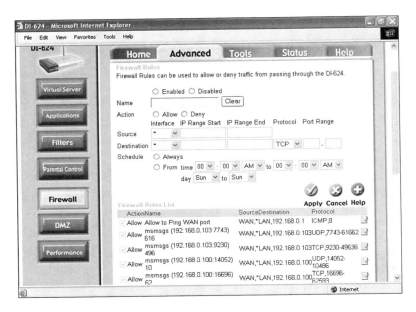

FIGURE 18.3

The firewall built in to many access points can be used to allow or prevent traffic from passing through the device.

With most home or small office wireless routers, a network firewall is implemented automatically. For the most part, you don't need to do anything to administer it. The exception to this is that sometimes specific applications require inbound or outbound access through one of the (many) ports blocked by the firewall (for example, special-ized content management servers). You can use the Firewall Rules screen to "punch a hole" in your network firewall, although obviously this is something to be done with care and only if you know what you are doing.

Setting Up a VPN

If you are connecting to a home network using a public Wi-Fi hotspot, using a virtual private network (VPN), which acts as a kind of tunnel through the Internet, is a great way to enhance security.

Earlier in this chapter, I explained that using a VPN to isolate the Wi-Fi access point from the rest of the network and to restrict access to authorized users, is a great way to beef up network security.

You can buy dedicated remote access servers that provide VPN functionality. For example, the Watchguard SoHo Firebox that I mentioned earlier is a good dedicated box for the SoHo class network that provides firewall and VPN capabilities. You can also buy sophisticated software to enable a VPN.

But why pay for it if it is available free? Windows XP Professional already includes a VPN remote access server.

To set up your VPN using Windows XP Professional, open the Network Connections window by clicking on Network Connections in the Control panel. Next, click Create a New Connection in the Tasks pane on the upper left of the Network Connections window.

The New Connection Wizard will open with a Welcome screen. Click Next to get started. In the Network Connection Type pane of the Wizard, choose Set Up an Advanced Connection as shown in Figure 18.4.

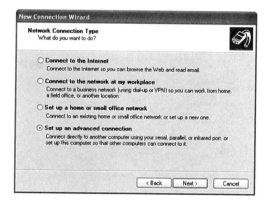

FIGURE 18.4

Choose Set Up an Advanced Connection to create a VPN in Windows XP.

Click Next. In the Advanced Connection Options pane, choose Accept Incoming Connections as shown in Figure 18.5.

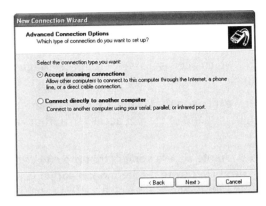

FIGURE 18.5

A VPN server should be set to accept incoming connections, or what is the point?

Click Next. The Devices for Incoming Connections pane will probably show your parallel port (LPT1) and nothing else. Don't do anything in this pane. Just click Next to continue setting up your VPN server.

In the Incoming VPN Connection pane, choose Allow Virtual Private Connections.

Click Next. In the User Permissions pane, shown in Figure 18.6, you can specify the users who have permission to use the VPN.

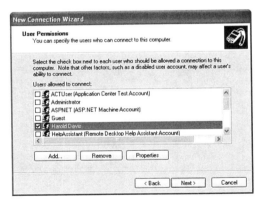

FIGURE 18.6

In the User Permissions pane, specify the users who can use the VPN.

There are a number of good features in specifying the users who can use the VPN in this way. First of all, access to the VPN is authenticated using the authentication controls baked in to the operating system. Second, users who access the VPN have only the privileges on the network that they've been granted. So guests, for example, might only have the right to read certain files (and no right to delete files).

Click Next. The Networking Software pane, shown in Figure 18.7, will open.

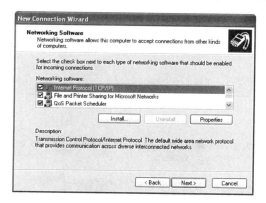

FIGURE 18.7

Select the networking software that should be enabled for incoming connections.

In the Networking Software pane, with the Internet Protocol (TCP/IP) item selected, click Properties. In the Incoming TCP/IP Properties window, shown in Figure 18.8, determine whether IP addresses for VPN clients, or callers, should be assigned by DHCP or provide a scheme for IP assignment.

Click OK to close the Incoming TCP/IP Properties window. Click Next to move to the final Wizard pane. Click Finish to create the VPN server, which will now be shown as an incoming connection in the Network Connections window, as you can see in Figure 18.9.

> **TIP**
>
> Now that the VPN server has been added as an incoming connection, you can edit it by selecting it in the Network Connections window and choosing Properties from its context menu. You don't have to run the New Connection Wizard again.

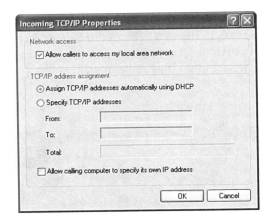

FIGURE 18.8

In the Incoming TCP/IP Properties window, choose to have IP addresses assigned using DHCP or designate an IP addressing scheme.

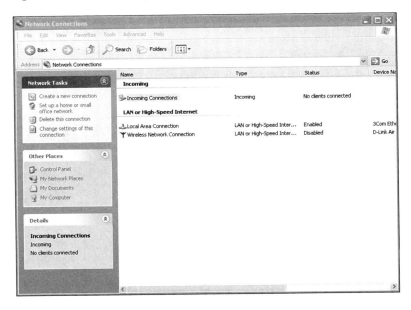

FIGURE 18.9

The VPN server is shown as an incoming connection in the Network Connections window.

Summary

Here are the key points to remember from this chapter:

- Because Wi-Fi networks are not physically secure, some level of security protection is a good idea.

- The level of security protection that you require depends on the confidentiality of the information you are protecting.

- If you do nothing else, you should change the default SSID for your wireless access point, set it not to broadcast the SSID, set the access point to use WEP encryption, and change the default administrative password for the access point.

- Complete security for a wireless network is probably impossible, but there are many steps you can take to make your Wi-Fi network more secure.

> **NOTE**
>
> If the VPN is behind a router, as will often be the case, for this setup to work, the router will have to be configured to automatically forward communications from the appropriate ports to the VPN server, a process called *port mapping*. The ports used for VPN access are forwarded to the IP for the VPN server.
>
> The ports used for VPN access depend on the VPN protocol used. Point-to-Point Tunneling Protocol (PPTP) uses ports 47 and 1723. Layer Two Tunneling Protocol (L2TP) uses ports 50, 51, and 500.

APPENDIXES

PART VI

Wireless Standards

Wi-Fi is a certified version of variants of the 802.11 wireless standards developed by the IEEE (http://www.ieee.org). Wi-Fi equipment is certified by the Wi-Fi Alliance (http://www.wi-fi.org) for compatibility and standards compliance.

The 802.11 wireless standards, also called *protocols*, are designed to use the so-called *free spectrums*, which do not require specific licensing. The spectrums currently used by 802.11 are 2.4GHz and 5GHz. (The 2.4GHz spectrum is also used by household appliances such as cordless phones. Microwave ovens also generate noise/interference in this radio spectrum, which can impact the performance of wireless devices.)

You don't need to know or give a fig about wireless standards or 802.11 to happily use wireless networking on the road or in your home network. That's why I've kept this material out of the body of *Anywhere Computing with Laptops: Making Mobile Easier*—because you don't really need to know it unless you are practicing to be an engineer.

But in case you are curious, I've summarized the most important concepts that underlie 802.11 (and Wi-Fi) in this appendix. Oh, and if you go into a store and a salesperson starts spouting "802.11 this and 802.11 that," you can check the summary of the 802.11 variants in "The Flavors of 802.11" and know exactly what is being talked about (quite likely better than the salesperson).

Understanding 802.11 Wireless Standards

The IEEE (Institute of Electrical and Electronics Engineers) Standards Association, http://standards.ieee.org/, likes to designate standards using numbers rather than names. Within the IEEE schema, the number 802 is used to designate local area networks and metropolitan, or wide area, networks (LANs and WANs). 802.11 is the name for wireless LAN specifications in general. There are a number of different flavors of the 802.11 standards, such as 802.11b, 802.11g, and so on, that I'll discuss later in this appendix in more detail. Each of these particular versions of the wireless LAN specification (802.11) run on the 2.4GHz or the 5GHz spectrums at high speeds.

To comprehend 802.11, it's important to understand the purpose of the general wireless LAN standard. According to the original Project Authorization Request, "the scope of the proposed standard is to develop a specification for wireless connectivity for fixed, portable, and moving stations within a local area." In addition, "...the purpose of the standard is to provide wireless connectivity to automatic machinery and equipment or stations that require rapid deployment, which may be portable, handheld, or which may be mounted on moving vehicles within a local area."

The 802.11 standards specify a *Medium Access Control* (MAC) layer and a *Physical* (PHY) layer using *Direct Sequence Spread Spectrum* (DSSS) technology.

The MAC layer is a set of protocols responsible for maintaining order in the use of a shared medium. For example, data encryption is handled in the MAC.

The PHY layer handles transmission between nodes (or devices on the network such as computers or printers). In other words, it is primarily concerned with hardware.

DSSS is a technique for splitting up and recombining information to prevent collisions between different data streams. Effectively, if my network is using DSSS and encounters other signals on the same spectrum it is using, my network can use DSSS to avoid interference from the other signals.

The MAC and PHY layers fit within the generalized OSI (Open System Interconnection) reference model. The OSI model is a way of describing how different applications and protocols interact on network-aware devices.

The primary purpose of 802.11 is to deliver MAC Service Data Units ("MSDUs") between Logical Link Controls ("LLCs"). Essentially, an LLC is a base station with a wireless access point, which itself might be connected to a wire line network for handoff to additional wireless LLCs.

802.11 networks operate in one of two modes: "infrastructure" and "ad hoc." The infrastructure architecture is used to provide network communications between wireless clients and wired network resources. An ad hoc network architecture is used to support mutual communication between wireless clients. It is typically created spontaneously, does not support access to wired networks, and does not require an access point to be part of the network.

The PHY layer of 802.11 defines three physical characteristics for wireless local area networks:

- Diffused infrared

- Direct Sequence Spread Spectrum (the primary technique for avoiding signal interference in today's 802.11 networks)

- Frequency Hopping Spread Spectrum (which is another technique for avoiding interference)

Okay! Enough jargon, acronyms, and theoretical engineering. Let's get on to the flavors of 802.11.

The Flavors of 802.11

This section briefly describes the variants of the 802.11 standard.

802.11b

Many Wi-Fi devices currently in operation are using the 802.11b flavor of 802.11 (although the backward compatible and faster 802.11g has replaced 802.11b as the reigning "monarch" of Wi-Fi).

The full 802.11b specification document is more than 500 pages long, but the most important things to know about 802.11b are that

- 802.11b uses the 2.4GHz spectrum.

- 802.11b has a theoretical throughput speed of 11 megabytes per second (Mbps).

> **NOTE**
> It's perfectly possible for a wireless chip to run multiple flavors of 802.11. For example, one of the chipsets available to laptops using Intel Centrino mobile technology runs 802.11b, 802.11g, and 802.11a. This "three-way" Wi-Fi radio will choose the most appropriate 802.11 variant depending on the network it is connecting to and the preferences of the user.

An 11Mbps speed isn't bad. However, for a variety of reasons, Wi-Fi connections rarely achieve anything close to 11Mbps. Over an encrypted 802.11b connection, you'll be lucky to get transmission rates of more than 6Mbps. This is certainly fast enough for transferring Word documents, but probably not fast enough for applications such as streaming video.

802.11g

If you go out today and buy a wireless laptop, or a wireless access point, it is most likely to use 802.11g. Like 802.11b, 802.11g runs on the 2.4GHz spectrum. One of the best things about 802.11g is that it is fully backward compatible with 802.11b. If your Wi-Fi laptop is equipped with 802.11b, you can connect to an 802.11g hotspot, although of course you will only achieve 802.11b throughput. (There are also some questions about whether placing 802.11b devices on an 802.11g network will degrade the entire network, not just the 802.11b portions of it, although this problem has mostly been resolved.)

Conversely, an 802.11g-equipped computer can connect to an 802.11b access point, once again at the lower speeds of 802.11b.

By comparison to 802.11b, 802.11g is blazingly fast, achieving throughput in the best conditions of 54Mbps. (Real-world actual rates tend to be around 25MBps.) This is still a little slower than a sophisticated 100Mbps wired Ethernet network but quite fast enough for most applications including streaming video and music. (Video and music streaming quality depends on many factors, such as lost packet rate, not just raw throughput.)

802.11a

Unlike either 802.11b or 802.11g, 802.11a operates on the 5GHz spectrum. That's good news from the viewpoint of interference. There's simply less going on in the 5GHz band, and you are less likely to "bump into" other Wi-Fi networks, garage door openers, cordless phones, or whatnot. However, it is bad news from the point of backward compatibility because 802.11a systems are not compatible with 802.11b (or 802.11g) because they use a different spectrum.

As I noted earlier, some vendors have solved this problem by creating tri-mode chipsets that run 802.11a, 802.11b, and 802.11g depending on the access point or hotspot they are connecting to. If you are considering purchasing an 802.11a Wi-Fi device, make sure that it has this kind of standard switching capability.

You can expect a throughput of something like 20Mbps with 802.a, so from a speed viewpoint, it is faster than 802.11b and pretty comparable 802.11g.

Generally speaking, you'd buy a Wi-Fi computer that only uses 802.11a if you are using it on a dedicated 802.11a network—for example, run by your company—and don't plan to take it anywhere else. However, the safest thing to do is to buy a trimode Wi-Fi computer that works with all three standards.

802.11n

802.11n is a developing standard not yet approved by the IEEE that promises to deliver greater than 100 Mbps (and perhaps as fast as 500Mbps) using both the 2.4GHz and the 5GHz spectrums. Despite the fact that the official standards approval process is not complete, some companies, such as Belkin and Broadcom, are likely to jump the gun and start producing "pre-standards approval" 802.11n chipsets and devices shortly.

Complicating matters, at the time of this writing, there are two different 802.11n camps, with different ideas about what the 802.11n standard should be when it is finalized.

THE COMPETING 802.11N STANDARDS

TGn Sync (Task Group N Sync) members include Nokia, Samsung, and Intel. The proposed TGn Sync 802.11n standard boasts high speeds and adaptive radio technology, enabling it to adapt to differing data streams and varying amounts of spectrum.

The competing group is known as WWiSE (short for *World Wide Spectrum Efficiency*). WWiSE members include Broadcom, Conexant, STMicroelectronics, and Texas Instruments. The version of 802.11n proposed by WWiSE is somewhat slower than the TGn Sync version, but proposes better backward compatibility and interoperability with the older Wi-Fi standards, such as 802.11g and 802.11b.

Still, in the due fullness of time, and when the dust settles, no doubt a workable 802.11n Wi-Fi standard will emerge. In effect, 802.11n is part of a process of better technology becoming more affordable. Right now, 802.11b is downright cheap. Equipment made using the 802.11g standard is faster, and on the market—but a bit more expensive. It won't be long before 802.11g is inexpensive, with new, faster (but higher priced) 802.11n equipment coming on the market. Further down the time horizon, we can expect the same process to be repeated with 802.16, commonly known as WiMAX.

802.11i

802.11i is the name given by the IEEE to its security standard. This standard is virtually the same as the "Wi-Fi Protected Access," or WPA2, the version adopted by the Wi-Fi Alliance.

Where the Hotspots Are

IN THIS APPENDIX
- Online directories
- Wi-Fi networks
- Retail locations
- Hotels
- Airports
- Free public networks

This appendix lists online directories of hotspots, provides information about national Wi-Fi networks, and tells you where you can find hotspots in retails chains, hotels, and airports. It also includes information about free, public access hotspots put up by libraries and some municipalities.

The problem is that, as they say, "you can't get there from here": An online source of information doesn't do you any good if you are not already online.

Besides online directories and Wi-Fi networks, this appendix lists the places where you are most likely to find a public Wi-Fi hotspot. Some of these are well known; for example, almost everyone knows that you can find a Wi-Fi hotspot at many Starbucks coffee shops. Others might surprise you.

Of course, no such listing can be complete. It would take an entire book similar to a phone directory. Here are some mind-boggling statistics: More than 25,000 hotspots are located in the United States alone. And, if you enter the phrase "Wi-Fi hotspots" into the Google search engine, it will return more than 500,000 hits.

So the problem is as much about sorting through this sea of hotspots and information to locate the hotspot you need when you need it as it is about simply finding a hotspot.

Also, the number of hotspots is expanding and changing every day. But getting accurate information about the location of hotspots can be difficult, so this appendix is intended to give you a "leg up" on the task.

The categories I've included are "Retail Locations," "Hotels," "Airports," and "Free Public Networks"—meaning public non-commercial Wi-Fi hotspots that you don't have to pay for.

Online Directories

Provided you have online access, far and away the best way to find Wi-Fi hotspots is an online directory. The hitch, of course, is that you have to be online in the first place. These directories don't do you much good if you are wandering aimlessly around "Podunk" with your Wi-Fi laptop or PDA all ready to go. The moral is do your research before you leave on your travels.

This section lists the best online directories. You can use them to find hotspots of all sorts (both free and for pay) in all locations.

- China Pulse: `http://www.chinapulse.com/wifi` (hotspot locater for China)
- HotSpotList: `http://www.wi-fihotspotlist.com`
- i-Spot Access: `http://www.i-spotaccess.com/directory.asp` (currently limited to Illinois, Iowa, Missouri, and Nebraska)
- Intel's Hotspot Finder: `http://intel.jiwire.com` (Features hotspots that have been verified for Intel Centrino mobile technology by Intel's Wireless Verification Program. Intel Centrino-verified hotspot information also appears on relevant Yahoo! maps.)
- JIWIRE: `http://www.jiwire.com`
- MetroFreeFi: `http://metrofreefi.com/`
- Ordnance Survey: `http://www.ordnancesurvey.co.uk/oswebsite/business/sectors/wireless/wifihotspot.html` (great for United Kingdom hotspots)
- Square 7: `http://www.square7.com/btopenzone/directory.htm` (great for European hotspots)
- WiFi411: `http://www.wifi411.com`
- Wi-Fi-Freespot Directory: `http://www.wififreespot.com`
- WiFinder: `http://www.wifinder.com` (one of the best all-around international hotspot directories)
- WiFiMaps: `http://www.wifimaps.com`
- WiFiZone: `http://www.wi-fizone.org`
- Wireless Access List: `http://www.ezgoal.com/hotspots/` (categorized by state and ZIP Code, also allows sorting by network; for example, T-Mobile, Wayport, and so on)

Wi-Fi Networks

Because Wi-Fi is such a new field, networks of Wi-Fi hotspots are still relatively small. One implication is that incumbent telecommunication providers—such as Cingular, Sprint, T-Mobile, and Verizon—have a head start.

But I've also included "mom-and-pop" networks. After all, the sheer number of hotspots might not be what counts for you: You might be most concerned with particular locations. I've tried to include every commercial Wi-Fi network that currently operates 50 or more hotspots.

From a user's point of view, you should check the availability and cost of roaming features before you sign up with a Wi-Fi network—because no one network is likely to provide all the hotspots you want to use. See Chapter 10, "Working with National Wi-Fi Networks," for more tips and techniques related to working with Wi-Fi networks.

Table B.1 shows the biggest networks as of the date of this writing, their Web address and telephone contacts, and current number of hotspots.

> **NOTE**
> The hotspot counts in this section are based on the actual number of hotspots that I could locate using the directories listed earlier in this appendix. The number of hotspots that I've noted are probably low now, and there will certainly be more by the time you read this book, but they do give an idea of the relative size of these networks.
> You should know that most of the Wi-Fi networks claim a greater number of hotspots than I've shown in this section, but I'm sticking to the number that I can actually verify.

The contact information in this table might be useful when you decide which network to sign up with. It could also help if you are traveling to an area that is particularly well served by a specific provider.

Table B.1 Wi-Fi Networks and Contact Info

Network	URL	Phone	Comments
Cingular Wireless	`http://www.cingular.com sbusiness/wifi/`	888-290-4613	Cingular's Wi-Fi service is formerly AT&T.
Boingo Wireless	`http://www.boingo.com`	800-880-4117	A pioneer Wi-Fi network, and still considered one of the best. Great roaming policies.

Table B.1 Continued

Network	URL	Phone	Comments
Sprint PCS	`http://www.sprint.com/ pcsbusiness/products_ services/data/wifi/ index.html` or `http://www.wifi.sprintpcs.com`	866-727-9434	Rolling out quickly based on existing infrastructure.
T-Mobile Hotspot	`http://www.t-mobile.com/ hotspot/`	877-822-7768	The leading Wi-Fi provider, building infrastructure in Starbucks stores and elsewhere. T-Mobile is the one everyone else wants to beat.
Wayport	`http://www.wayport.com`	888-929-7678	An early pioneer in providing Wi-Fi hotspots, now providing infra-structure in McDonald's restaurants and elsewhere.

Retail Locations

Here are some national chains that provide free Wi-Fi access in some or all of their stores:

- All Apple retail stores. See `http://www.apple.com/retail/` for locations.

- Panera Bread is aiming to have all its locations offer free Wi-Fi. In the meantime, many Panera stores already are "unwired." See `http://www.panera.com/ wifi.aspx` for details.

- Many Schlotzsky's Deli locations. See `http://www.schlotzskys.com/wireless. html` for a list.

Here are a few of the national chains that provide Wi-Fi access in some or all of their stores using a paid provider (this means that you have to pay for your access):

- Barnes & Noble stores rely on various different providers. The best way to find a Barnes & Noble location that provides Wi-Fi access is to use one of the directories listed in the section "Online Directories."

- Borders bookstores are "unwired" using T-Mobile Hotspot's network. The best way to find a Borders location that provides Wi-Fi access is to use one of the directories listed in the section "Online Directories."

- All FedEx Kinko's stores offer Wi-Fi access through T-Mobile Hotspot. See `http://www.fedex.com/us/officeprint/storesvcs/technology/wifi.html` for information and locations.

- Some McDonald's locations are being equipped with Wi-Fi access. The providers vary depending on the location, but usually it is Wayport. See `http://www.mcdonalds.com/wireless.html` for more information.

- Starbucks Wi-Fi access is provided by T-Mobile Hotspot. See `http://www.starbucks.com/retail/wireless.asp` for locations with this service.

Hotels

Many hotel chains offer Wi-Fi access on a paid or free basis in both hotel rooms and the hotel lobby. Wi-Fi is available in some or all the facilities of a given hotel chain, so be sure to check with the hotel in which you will be staying to see if Wi-Fi is available.

Hotel chains with free service include

- Best Western
- Comfort Suites and Clarion
- Courtyard by Marriott
- Holiday Inn Express
- Omni Hotels

Here are some hotel chains that provide paid Wi-Fi access:

- Doubletree
- Hilton
- Hyatt
- Loews
- Marriott
- Radisson
- Sheraton
- W
- Westin
- Wyndham

Check with each individual hotel facility to determine whether Wi-Fi is in fact available, the actual cost, and the network provider (if any).

Airports

Paid Wi-Fi access is available in many major airports, catering to stranded travelers. This kind of Wi-Fi service can be provided in special airline lounges, or it can be more widely available throughout passenger terminals.

Table B.2 shows some of the major American airports with widespread Wi-Fi access.

Table B.2   Airports with Wi-Fi

Airport	Network Provider	Airport	Network Provider
Atlanta (ATL)	T-Mobile Hotspot	Newark (EWJ)	Telia Homerun
Chicago (ORD)	T-Mobile Hotspot	Oakland (OAK)	Wayport
Dallas (DFW)	T-Mobile, Wayport	Pittsburgh (PIT)	Free Access
Fort Lauderdale-Hollywood (FLL)	Free access in most of the Croward County, Florida airport	Rochester, NY (ROC)	Free access in airport waiting areas
John F Kennedy, New York (JFK)	Free access at JetBlue gates	San Francisco (SFP)	T-Mobile Hotspot
Los Angeles (LAX)	The Gate Escape, a company with several facilities in LAX that rents office services to airport travelers	San Jose (SJC)	Wayport
		Delta Crown Rooms (airline lounge)	T-Mobile hotspot

Free Public Networks

As I noted in the listing of Wi-Fi directories, Wi-Fi-Freespot Directory, http://www.wififreespot.com, focuses on free hotspots. This is a good starting place if you are looking for no-cost Wi-Fi public access. It is also worth noting that some of the big telecom providers give free Wi-Fi access to users subscribing to some of their other services.

Most of the free hotspots fall into one of three categories:

- Provided by a business in the hope of drawing traffic

- Provided by a library as a public service (see the following discussion)

- Provided by a local community (see the following discussion)

It's really worth browsing Wi-Fi-Freespot to see what local businesses you can find that provide free Wi-Fi. You'll probably be absolutely amazed at how many there are. (I know; I know: My community, Berkeley, California, is not typical of anything, but there are literally hundreds here.)

Many public libraries provide Wi-Fi free access to the public (or at least to members of the community who have registered for a library card). The Wireless Librarian has a great directory of Wi-Fi availability in libraries: `http://people.morrisville.edu/ ~drewwe/wireless/wirelesslibraries.htm`.

Wireless Communities, `http://wiki.personaltelco.net/index.cgi/ WirelessCommunities`, provides information about community-maintained Wi-Fi hotspots all over the world. This is worth checking out. You'll be surprised at how many community-based providers of free Wi-Fi networks there are, as well as the passion and commitment of the people and communities that provide this free public service.

Many municipalities, such as Austin, Texas and Philadelphia, Pennsylvania, are rushing to put up municipal Wi-Fi hotspot networks. As these go up, here are a few sample wireless communities of relatively long standing:

- Bay Area Wireless Users Group (BAWUG) is one of the pioneering free Wi-Fi organizations. In addition to BAWUG's other activities, the group maintains a number of free hotspots centered around the Presidio in San Francisco. `http://www.bawug.org`

- Detroit Wireless Project (DWP) helps maintain a number of free hotspots in downtown Detroit. `http://www.dwp.org`

- Personal Telco Project (PTP) has more than 100 hotspots in Portland, Oregon, and has contributed to Portland being named the "hottest" city—or city with the most hotspots—in the United States for several years running. `http://www. personaltelco.net/static/index.html`

- Seattle Wireless, in addition to supported hotspots around the Seattle area, has pioneered an alternative public backbone for wireless networking. `http://www. seattlewireless.net`

WI-FI TO THE MAX: TAKE YOUR MOBILE COMPUTER TO SCHOOL!

Almost every university and college in the United States offers free Wi-Fi access to its students. This access is not, in most cases, available to the general public. However, if you are attending college, or plan to start next year, it is very likely that your school will provide convenient Wi-Fi access for you. You should check with your school to learn the details of the Wi-Fi access provided and plan to bring your laptop.

In addition, it is often possible to arrange for guest access to a university or college wireless network, particularly if you will be using it for scholarly research.

Intel Centrino Mobile Technology Platform

ntel Centrino mobile technology is a platform incorporating Intel's technologies designed specifically for mobile computing with integrated wireless LAN capability. The Intel Centrino mobile technology platform combines a central processing unit (CPU) and chipset designed for mobile computing with wireless LAN functionality using the Wi-Fi standards. The great bulk of today's laptop computers are built using Intel Centrino mobile technology.

If you are like most computer users, this might be all you really want or need to know about the Intel Centrino mobile technology platform. If so, you can skip this appendix, which provides some more detailed information about the Intel Centrino mobile technology platform. But read on if you'd like to know more.

Product Information and Components

Three in one and one for all! Designed from the ground up for anywhere computing with integrated wireless LAN capability, breakthrough mobile performance while enabling extended battery life and sleek, intended for easy-to-carry notebook PCs, Intel Centrino mobile technology allows users to stay connected and productive on the go—for a truly happy, mobile lifestyle.

It's like an old-fashioned Chinese restaurant menu. Each laptop built using the Intel Centrino mobile technology platform has one of each of these three components:

- An Intel Pentium M processor (or CPU)

- A member of the Intel 855 or Intel 915 Express chipset family, used for input/output (i/o) processing

- A member of the Intel PRO/Wireless Network Connection family

TIP

You'll find much more information about the Intel Centrino mobile technology platform at http://www.intel.com/products/centrino/index.htm.

With Intel Centrino mobile technology, three components work together to enable outstanding mobile performance, extended battery life, and integrated wireless LAN capability in thinner and lighter notebooks. As a whole, the platform delivers mobile performance and low-power enhancements for sleeker, lighter notebook designs.

Processor and Chipset

This section looks in more detail at two of the components of the Intel Centrino mobile technology platform—the Intel Pentium M processor and the surrounding Intel chipsets (855 and 915).

Intel Pentium M Processor

The Intel Pentium M provides many advanced features including

- A micro architecture including power-optimized 400MHz or 533MHz processor system bus, Micro-Ops Fusion, and Dedicated Stack Manager for faster execution of instructions at lower power

- Normal, Low Voltage (LV), and Ultra Low Voltage (ULV) processor models to assist laptop designers in meeting their goals for performance and battery life

- Advanced instruction prediction to reduce re-dos for increased performance

- Support for Enhanced Intel SpeedStep technology with multiple voltage and frequency operating points, enabling a better match of performance to application demand

- Support for Intel Mobile Voltage Positioning, which dynamically lowers voltage based on processor activity to lower thermal design power enabling smaller notebooks

- Low power technologies

- 1MB power managed L2 cache increasing system performance

The Intel Pentium M is provided in many different models, with model numbers currently ranging from 715 to 770. These models feature 90nm and 130nm process technology, L2 cache, and clock speeds from 1.3–2.13GHz. The Pentium M series supports up to 2GB of DDR SDRAM or 2GB of DDR2 SDRAM, depending on the supporting chipset employed.

Intel 855 Chipset

The Intel 855 chipset family is part of Intel Centrino mobile technology and is designed to deliver excellent performance with lower demands for power (important in a mobile computer). There are a number of sub-chipsets that are part of the Intel 855 chipset family, including

- The Intel 855PM chipset is a discrete memory controller hub.

- The Intel 855GM chipset is an integrated graphics memory controller hub.

- The Intel 855GME chipset is a graphics memory controller hub featuring a highly integrated mobile solution optimized to support the Intel Pentium M processor.

Intel 855GM Chipset

The Intel 855GM Chipset graphics memory controller hub (GMCH-M) is a highly integrated mobile chipset solution that has been optimized to support the Intel Pentium M processor, high speed DDR memory, and a Hub interface. In addition to this, the Intel 855GM chipset provides integrated graphics capabilities.

Intel 855PM Chipset

The Intel 855PM Chipset memory controller hub (MCH-M) is a mobile chipset solution that has been optimized to support the Intel Pentium M processor, high speed DDR memory, and a Hub interface. This chipset has an AGP 4X interface and provides flexible support for high performance discrete graphics solutions.

Intel 855GME Chipset

The Intel 855GME chipset graphics memory controller hub (GMCH-M) is a component of Intel Centrino mobile technology. The chipset is a highly integrated mobile solution optimized to support the Intel Pentium M processor, high speed DDR memory, and a Hub interface.

Intel 915 Express Chipset

The Intel 915 Express chipset family is the second-generation Centrino supporting chipset with several important improvements over the original Intel 855 chipset family. These improvements include

- Support for Intel Pentium M processors with 533MHz Front Side Bus (FSB)

- Support for higher-speed DDR2 SDRAM

- Support for the new PCI Express standard

- Support for Serial ATA (SATA) as well as IDE disk drives

- Intel High-Definition Audio

- The Intel Graphics Media Accelerator 900 video controller

Intel Wireless Network Connection

The Intel Wireless Network Connection family—a little more formally, the Intel PRO/Wireless network connection family—is an integrated wireless solution that is central to the Intel Centrino mobile technology platform and is Wi-Fi certified.

In addition, the Intel Wireless Network Connection family supports most industry standard wireless LAN security standards. Enabling software for these security standards is available.

The Intel Wireless Network Connection family features Intel Intelligent Scanning Technology, which reduces power by controlling the frequency of scanning for access points.

The Intel PRO/Wireless network connection works in concert with the other Intel Centrino mobile technology components to provide freedom and flexibility to work and play on the go without hunting for a phone jack, network cable, or plugging in a special card.

The Intel Wireless Network Connection family provides deployment flexibility and connectivity convenience by offering a choice of products including a single band (supporting 802.11b), a dual band (supporting 802.11a and 802.11b), and a dual mode (supporting 802.11b and 802.11g) product (the most common version you'll find today). In addition, a new variant member of the Intel Wireless Connection family is tri-mode, supporting all the current versions of Wi-Fi (802.11a, 802.11b, and 802.11g).

With throughput up to 54Mbps at 5GHz (802.11a) and 2.4GHz (802.11g), the Intel Wireless Network Connection family enables fast wireless network connections. The Intel Wireless Coexistence System helps reduce interference with certain Bluetooth devices. Power Save Protocol (PSP) is a user selectable feature with five different power states, allowing the user to make his own power versus performance choice when in battery mode. This feature supports the extended battery life benefits of Intel Centrino mobile technology.

The typical maximum indoor range for a member of the Intel Wireless Connection family with an ordinary wireless access point without an added antenna is between 100 and 300 feet.

Platform Benefits and Vision

Wireless computing, anywhere computing, working on the go: You know the drill and the benefits. It's great to surf while you sip, to surf poolside, to work while you're stuck in an airport or hotel room, and to be able to set up networks in you home or small business without having to bother with stringing network cables. Myself, I really like to take my Intel Centrino laptop out to my garden and work in the glorious sunshine!

Intel points to a number of more tangible benefits for the Intel Centrino mobile technology platform, including the reduction of capital costs for businesses that use wireless technology, reduced need for technical support, and less user downtime.

Not only does integrating wireless local area network (WLAN) capability save up-front capital costs, but also it reduces costs because of damaged wireless cards (which are often lost, spindled, or mutilated).

Because Intel has worked hard to make sure that the Intel Centrino mobile technology platform is compatible with access points made by almost every leading vendor, you are less likely to have major technical support issues with the platform than with wireless equipment that is not fully integrated into your computing platform. The wireless mobile computer form factor represented by the Intel Centrino mobile technology platform should lead to a reduction in administrative costs as well.

The extensive validation and testing of Intel Centrino mobile technology includes the CPU, chipset, and wireless components and continues as an ongoing matter. In addition, Intel is working with leading WLAN security solution vendors to test and validate Intel Centrino mobile technology with their advanced security products, and ultimately to design more secure wireless solutions.

The vision of the Intel Centrino mobile technology platform is

- Anywhere, anytime computing, and connectivity

- Easy, seamless, and reliable wireless connectivity

- Peace, universal happiness, and a more productive world population (only kidding!)

- Secure and safe computing with wireless connectivity

- Lightweight and thin mobile computers that make anywhere computing fun and easy

- Longer battery life, making anywhere computing more convenient

- Excellent computing performance to complement easy anywhere connectivity

Glossary

10-BASE T Ethernet   Standard for wired networks that use phone-like plugs.

3G   A catchall term used to refer to a proprietary high-speed wireless network using spectrums leased by telecommunications carriers that is proposed and/or has partially been developed.

802.11   The general standard for wireless networking, defined by the IEEE.

802.11a   A relatively new version of the 802.11 wireless standard that is faster than 802.11b (it runs at speeds up to 25Mbps) and uses the 5GHz spectrum. 802.11a is not backward compatible with 802.11b, the older flavor of Wi-Fi.

802.11b   Original flavor of 802.11 wireless networking, or Wi-Fi. 802.11b uses the 2.4GHz spectrum and has a theoretical speed of 11 Mbps.

802.11g   The current most popular version of the 802.11 wireless standard, both newer and faster than 802.11b. 802.11g runs on the 2.4GHz spectrum, is backward compatible with 802.11b, and has a theoretical speed of 54Mbps.

802.11i   The name given by the IEEE to the security standard requiring encryption of transmissions, substantially similar to the "Wi-Fi Protected Access," or WPA2 for short, standard promoted by the Wi-Fi Alliance.

802.11n   A new 802.11 wireless standard that promises to deliver speeds of well over 100Mbps using both the 2.4GHz and 5GHz spectrums. So far, 802.11n has not yet received IEEE approval, and the details of the standard are the subject of industry controversy.

802.16   The IEEE standard designation for the standard used by WiMAX, a wireless technology that promises much greater bandwidth than Wi-Fi and 802.11.

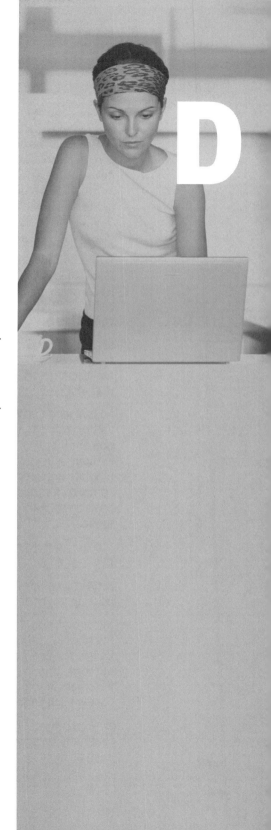

access control layer   Baked into the Wi-Fi standards; specifies how a Wi-Fi device, such as a mobile computer, communicates with another Wi-Fi device, such as a wireless access point, and specifies a unique hardware identifier for each network device. Also called MAC layer.

access point   A broadcast station that Wi-Fi computers can communicate with. Also called AP, hotspot, and base station. Access points are used as the central point for a network of Wi-Fi computers.

access point/router   This common hardware combination combines the functionality of a Wi-Fi access point with that of a network router.

ad hoc mode   Wi-Fi computers in the ad hoc mode communicate directly with one another in a peer-to-peer fashion without using an access point to intermediate network communication. (In contrast, a wireless network with a central access point uses *infrastructure* mode.)

AirPort   Apple's name for the 802.11b flavor of Wi-Fi.

AirPort Extreme   Apple's name for the 802.11g flavor of Wi-Fi.

authentication   The process of ensuring that an individual is who he or she claims to be; note that (as opposed to authorization) authentication does not automatically confer any rights to assets such as files.

authorization   The security process that handles access rights to assets such as files.

Bluetooth   A short-range connectivity solution designed for data exchange between devices such as printers, cell phones, and PDAs that use the 2.4GHz spectrum. Bluetooth is incompatible (meaning, does not work) with any of the 802.11 wireless networking standards.

bridge   Wireless bridges are used to connect one part of a network with another. They are often used to extend the range of a network and/or to add non-PC devices to a network. Wireless bridges can be set up using dedicated hardware; however, they can also be configured using the software in your Centrino laptop.

broadband   Fast delivery of Internet services. The predominant means for the delivery of broadband are cable and DSL, although if 802.11n and WiMAX deliver on their promises, they will become very viable broadband delivery technologies.

cable   Broadband Internet delivery over the same cable used to bring in television content, primarily to residential subscribers.

Centrino   *See* Intel Centrino mobile technology.

chipset   A chipset is a group of integrated circuits that can be used together to serve a single function and are therefore manufactured and sold as a unit. The functions that make up the core logic of a computer or other electronic device are often performed by a chipset.

client   A device or program connected to a server of some sort. Typically, a client is a personal computer connected to a server computer that relies on the server to perform some operations.

client-server network   In a client-server network, a centralized server computer controls and polices many of the basic functions of the network, and intermediates the communications between the computers on the network.

closed node   A Wi-Fi broadcast that is protected by WEP.

CompactFlash (CF) card   Used to add memory (and in some cases other functionality, such as Wi-Fi connectivity) to PDAs, digital cameras, and other handheld devices.

DDR   Double Data Rate (DDR) is a type of SDRAM in which data is sent on both the rising and falling edges of clock cycles in a data burst.

DHCP   Dynamic Host Configuration Protocol, a standard for assigning dynamic Internet Protocol (IP) addresses to devices on a network.

Direct Sequence Spread Spectrum (DSSS)   Technology used to prevent collisions and avoid interference between devices operating on the same wireless spectrum.

directional antenna   An antenna used to focus a transmission in a specific direction, used in point-to-point Wi-Fi broadcasts, also called a yagi.

DMZ   Demilitarized zone, an isolated computer or subnetwork that sits between an internal network that needs to remain secure and an area that allows external access.

DNS   The Domain Name System (sometimes called Domain Name Service) translates more or less alphabetic domain names into IP addresses.

DSL   Digital Subscriber Line, a technology used to deliver broadband Internet services over telephone lines.

dumpster diving   Going through trash to find personal information, often with the intention of obtaining identifying information and passwords to gain access to online assets.

dynamic IP addressing   The provision of IP addresses to computers dynamically.

EIRP   Unit of measure (equivalent isotropically radiated power) for the strength of an antenna.

encryption   The translation of data using a secret code; used by WEP to achieve security for Wi-Fi networks.

encryption key   The key used to encrypt the transmissions of a Wi-Fi network protected by WEP; the password needed to access a Wi-Fi network.

Ethernet port   A socket that accepts a 10BASE-T Ethernet wire.

FCC   Federal Communications Commission. *See* http://www.fcc.gov for more information.

369

firewall   A firewall is a blocking mechanism—either hardware, software, or both—that blocks intruders from accessing a network or an individual computer; an intermediary program that can be configured to control access both to and from a private network.

free radio spectrum   Bands of the radio spectrum, such as 2.4GHz and 5GHz, that do not require a license from the FCC. Wi-Fi uses these bands for its transmissions, as do other household appliances such as cordless phones and microwaves.

frequency   The oscillations, or movement from peak to trough, of the electromagnetic wave created by a radio transmission.

gain   The amount of gain provided by an antenna means how much it increases the power of a signal passed to it by the radio transmitter.

GHz   Gigahertz. One gigahertz equals 1,000 megahertz (MHz).

hotspot   An area in which Wi-Fi users can connect to the Internet, whether for pay or free.

hub   A simple wired device used to connect computers on a network.

IEEE   Institute of Electrical and Electronics Engineers. *See* http://www.ieee.org for more information.

infrastructure mode   A Wi-Fi network in infrastructure mode uses an access point to intermediate network traffic (as opposed to ad hoc mode, which features direct peer-to-peer communication).

Intel Centrino   *See* Intel Centrino mobile technology.

Intel Centrino mobile technology   A platform incorporating Intel's technologies designed specifically for mobile computing with integrated wireless LAN capability. It combines a central processing unit (CPU) and chipset designed for mobile computing with wireless LAN functionality using the Wi-Fi standards. The great bulk of today's laptop computers are built using Intel Centrino mobile technology.

intruder   Someone who gains illicit access to a private network.

intrusion   An incident involving illicit access to a network by an intruder.

IP address   A hexadecimal tuplet that denotes a node on the Internet or other network.

isotropic radiation pattern   If an antenna has an isotropic radiation pattern, the antenna transmits radio waves in all three dimensions equally.

ISP   Internet service provider, such as the cable or phone company providing you Internet access using dial-up, cable, or DSL.

IT   Information technology department, usually in a large enterprise.

key   Used to encrypt transmissions on a protected wireless network.

LAN   Local area network, such as the network in your home or small office.

latte   The beverage of choice to sip at a coffee shop while you are connected to the Internet via Wi-Fi.

logical topology   The logical data flow on a network.

MAC address   Media access control layer address; unique identification number of each network device.

MAC filtering   Creating a secure Wi-Fi network by using the MAC address of each device on the network (and only allowing devices with a known MAC address).

MAC layer   Media access control layer. *See* access control layer.

Mbps   Megabytes per second.

Media access control layer   *See* access control layer.

MHz   Megahertz. In a radio spectrum, 1 megahertz means one million vibrations per second.

modem   Short for modulator-demodulator. A modem is a device that lets a computer transmit data over telephone or cable lines and connects your cable or DSL Internet service to your computer or your network.

network   Two or more connected computers.

Network Address Translation (NAT) Translates local network addresses to ones that work on the Internet.

network name   *See* SSID.

NIC   Network interface card, used to connect a computer to a wired network.

omnidirectional antenna   An antenna that sends broadcast signals in all directions.

open node   A Wi-Fi broadcast that is not protected by WEP.

Open-System authentication   So-called Open-System authentication is used when no authentication is required because it does not provide authentication, only identification of the MAC address of a wireless client that is connected.

Open System Interconnection (OSI) A general reference model that describes how different applications and protocols interact on network-aware devices. The MAC and PHY layers of Wi-Fi compliant devices fit within the OSI model.

PC Card   Card that fits in the PCMCIA slot that is present on most laptops. Also called PCMCIA cards.

PCI Card   Card that fits into the PCI expansion slot inside a Windows desktop computer.

PCMCIA   Personal Computer Memory Card International Association, which is the name of the organization that has devised the standard for cards that can be added to laptops. For more information *see* http://www.pcmcia.org.

PCMCIA slot   Also called an expansion slot, used to add PC cards to a laptop.

371

PDA   Personal Digital Assistant, a handheld computer.

peer-to-peer network   In a peer-to-peer network, computers communicate directly with each other.

penetration   The act of illicitly accessing a private computer or network, usually performed by an intruder and resulting in an intrusion.

Pentium   An advanced class of CPUs for personal computers made by Intel Corporation.

Pentium M   A CPU of the Pentium class especially designed for mobile computers.

Physical layer (PHY)   Baked into the Wi-Fi standards; handles transmission between nodes (or devices) on a Wi-Fi network. In other words, it is primarily concerned with connectivity at the hardware level.

physical topology   The way a network is connected.

PPPoE   Point-to-Point Protocol over Ethernet, usually used with DSL for broadband connectivity to the Internet.

protocol   An agreed upon format for transmitting data between devices. *See also* standard.

radio bands   Contiguous portions of the radio spectrum—for example, the FM band.

radio spectrum   The entire set of radio frequencies.

Radius server   A server used for remote user authentication. Specifically, a Radius server is used to implement WPA authentication, authorization, and encryption on a wireless network using a key that changes automatically at a regular interval.

RAM   Random access memory.

random access memory (RAM)   Used to temporarily store instructions and information for the microprocessor of a computer or other device.

roaming   The ability to use a Wi-Fi network provider other than your Wi-Fi network provider through your original account.

router   A device that directs traffic between one network and another—for example, between the Internet and your home network.

SDRAM   Synchronous dynamic random access memory (SDRAM) is RAM that delivers bursts of data at high speeds using a synchronous interface.

server   A computer or device on a network that manages network resources.

Shared Key authentication   In Shared Key authentication, an access point sends out random bytes, which the wireless computer requesting access must encrypt using the shared key and send back to the access point. The access point then decrypts using the shared key and verifies that the result matches the original.

shoulder surfing   Reading confidential information on a computer over someone's shoulder.

social engineering   Tricking a person into revealing his password or other confidential information related to computers and networks.

SOHO   Small office or home office.

SSID   Service set identifier, used to identify the "station" broadcasting a Wi-Fi signal. Also called the network name, or wireless network name. Apple calls the SSID for its AirPort products the AirPort ID.

standard   Used in engineering to mean the technical form of something such as a message or a communication. *See also* protocol.

static IP address   An IP address that does not change.

switch   An intelligent hub used to connect computers on a network.

TCP/IP   Transmission Control Protocol/Internet Protocol, a protocol used by the Internet and other networks to standardize communications.

topology   A network topology is the arrangement of a network.

USB   Universal serial bus, used to connect peripheral devices such as a mouse to a computer.

VOIP   Voice over IP, a technology that enables telephone calls to be placed over the Internet.

VPN   Virtual private network, software used to "tunnel" through the Internet to provide secure access to remote resources.

WAN   Wide area network, such as the Internet.

WAP   Wireless application protocol, used to provide Internet capabilities, such as Web browsing, to "thin" wireless devices, such as mobile phones.

War chalking   The use of chalk markings on a sidewalk to identify Wi-Fi networks.

War driving   Cruising in a car with a Wi-Fi laptop looking for unprotected Wi-Fi networks.

WEP   Wired equivalent privacy, an encryption security standard built in to the current versions of Wi-Fi.

Wi-Fi   Short for *wireless fidelity*, is the Wi-Fi Alliance's name for a wireless standard, or protocol, used for wireless networking using the 802.11 standards.

Wi-Fi Alliance   A not-for-profit organization that certifies the interoperability of wireless devices built around the 802.11 standard. *See* http://www.wi-fi.org for more information.

Wi-Fi directory   A site on the Internet that provides tools to help you find Wi-Fi hotspots.

Wi-Fi finder   A device that senses the presence of active Wi-Fi networks.

Wi-Fi network provider   A company that provides access, usually for a fee, to multiple Wi-Fi hotspots.

Wi-Fi Protected Access   *See* 802.11i.

WiMAX   WiMAX is a new wireless technology (based on the IEEE 802.16 standard) that promises much greater bandwidth than Wi-Fi and 802.11. At the time of this writing, there are no commercial WiMAX deployments.

Wireless bridge   *See* bridge.

WPA   Wi-Fi Protected Access; *see also* 802.11i.

WPA-PSK   Wi-Fi Protected Access–Pre-Shared Key encryption means that the wireless computer and the access point must have the same pass phrase in order to set a wireless connection. WPA-PSK does not require a Radius server.

Yagi   *See* directional antenna.

INDEX

More Great Titles from Que Publishing

**Building a Digital Home Entertainment
Network: Multimedia in Every Room**
Terry Ulick
0-7897-3318-8
$24.99

**Family Computer Fun: Digital Ideas Using
Your Photos, Movies, and Music**
Ralph Bond
0-7897-3378-1
$24.99

Treo Essentials
Michael Morrison
0-7897-3328-5
$24.99

You Also Might Like

**Absolute Beginner's Guide
to Home Networking**
Mark Soper
0-7897-3205-X
$18.95

**Absolute Beginner's Guide
to WI-FI Wireless Networking**
Harold Davis
0-7897-3115-0
$18.95

**Absolute Beginner's Guide
to Home Automation**
Mark Soper
0-7897-3207-6
$21.99